Fisheries Management

Summer Walter

Callisto Reference

www.callistoreference.com

Callisto Reference,
118-35 Queens Blvd., Suite 400,
Forest Hills, NY 11375, USA

Visit us on the World Wide Web at:
www.callistoreference.com

ISBN: 978-1-64116-528-0 (Hardback)

Cataloging-in-Publication Data

Fisheries management / Summer Walter.
p. cm.
Includes bibliographical references and index.
ISBN 978-1-64116-528-0
1. Fishery management. 2. Fisheries. 3. Fishery co-management.
4. Fish management areas. I. Walter, Summer.

SH328 .F57 2022
338.372 7--dc23

TABLE OF CONTENTS

TABLE OF CONTENTS

PREFACE

This book is a culmination of my many years of practice in this field. I attribute the success of this book to my support group. I would like to thank my parents who have showered me with unconditional love and support and my peers and professors for their constant guidance.

Fisheries management is an activity that focuses on the protection of fishery resources to enable sustainable exploitation. It involves the application of concepts from fisheries science. The management of fisheries is an integrated process which includes planning, consultation, information gathering, analysis, allocation of resources as well as the formulation and implementation of regulations and rules which govern fishery activities. It seeks to ensure continued productivity of the resources and the accomplishment of other objectives of the fisheries. Some of the primary objectives of fisheries management are the maximizing of sustainable biomass yield, maximizing sustainable economic yield and increasing export income. This book elucidates the concepts and innovative models around prospective developments with respect to fisheries management. It picks up individual concepts and explains their need and contribution in the context of a growing economy. This book will serve as a valuable source of reference for those interested in this field.

The details of chapters are provided below for a progressive learning:

Chapter - What is a Fishery?

The industry which is involved in harvesting or raising fish is called fishery. Some of the different types of fisheries are fresh water fisheries and marine fisheries. This is an introductory chapter which will introduce briefly all these types of fisheries.

Chapter - Fish Farming

The raising of fish commercially in tanks or fish ponds usually for food is referred to as fish farming. It is also involved in the prevention of diseases in fish as well as their treatment. This chapter has been carefully written to provide an easy understanding of these facets of fish farming.

Chapter - Fishing Techniques

Various kinds of techniques are used in fishing. These include remote control fishing, gillnetting, longline fishing, seine fishing, trawling, electric pulse fishing, electrofishing, etc. This chapter closely examines these types of fishing techniques to provide an extensive understanding of the subject.

Chapter - Fishery Monitoring, Control and Surveillance

There are different kinds of systems used for monitoring and controlling a fishery. Vessel monitoring system, fishery resource monitoring system and catch reporting are a few parts of its monitoring and control system. The applications of these systems have been thoroughly discussed in this chapter.

Chapter - Fisheries Management

Fisheries management is the activity of protecting fishery resources in order to facilitate sustainable exploitation. It can be broadly categorized into conventional fishery management and ecosystem-based fisheries management. This chapter discusses in detail these types of fisheries management as well as the different fish diseases and parasites.

Chapter - Threats to Fisheries

Sustainable fishery refers to the harvesting of fishes at a sustainable rate such that fish population does not decline over time. Some of the challenges to sustainable fishery are overfishing, illegal fishing and marine pollution. The topics elaborated in this chapter will help in gaining a better perspective about sustainable fishery and its challenges.

Summer Walter

What is a Fishery?

CHAPTER 1

The industry which is involved in harvesting or raising fish is called fishery. Some of the different types of fisheries are fresh water fisheries and marine fisheries. This is an introductory chapter which will introduce briefly all these types of fisheries.

A fishery is an organized effort (industry, occupation) by humans to catch and/or process, normally for sale, fish, shellfish, or other aquatic organisms. The activity of catching the aquatic species is called fishing, and it is employed in the business of a fishery. Generally, a fishery exists for the purpose of providing human food, although other aims are possible, such as sport or recreational fishing), obtaining ornamental fish, or producing fish products such as fish oil. Industrial fisheries are fisheries where the catch is not intended for direct human consumption.

The focus of a fishery may be fish, but the definition is expanded to include shellfish (aquatic invertebrates such as mollusks, crustaceans, and echinoderms), cephalopods (mollusks, but sometimes not included in the definition of shellfish), and even amphibians (frogs), reptiles (turtles), and marine mammals (seals and whales, although "whaling" is the term usually used instead of fishing). Among common mollusks that are the target of a fishery are clams, mussels, oysters, and scallops, and such edible cephalopods as squid, octopus, and cuttlefish. Popular crustaceans are shrimp, prawns, lobsters, crabs, and crayfish, and representative echinoderms, which are popular in Asia, are sea cucumbers and sea urchins.

Fishing nets on a shrimp boat, Ostend.

Fisheries offer human beings the ability to utilize natural resources of the ocean and freshwater for human prosperity. However, people have created such efficient fishing technologies that people can now exploit nature to a capacity where the targeted species are endangered. Indeed, the history of human interactions with nature has often been one of exploitation for short-term gain, or with little regard for the impact on the animals exploited. As stewards of creation, human beings have the responsibility to regulate their harvesting of aquatic organisms to sustainable levels, where the prosperity of future human generations can be assured, including the aesthetic and non-commercial values offered by the species.

The fishing effort is generally centered on either a particular ecoregion or a particular species or type of fish or aquatic animal, and usually fisheries are differentiated by both criteria. Examples would be the salmon fishery of Alaska, the cod fishery off the Lofoten islands, or the tuna fishery of the Eastern Pacific. Most fisheries are marine, rather than freshwater; most marine fisheries are based near the coast. This is not only because harvesting from relatively shallow waters is easier than in the open ocean, but also because fish are much more abundant near the coastal shelf, due to coastal upwelling and the abundance of nutrients available there.

Importance

Fisheries have been important in various national economies throughout history, including the ancient Phoenicians, Greeks, and Egyptians. Spanish fisheries helped to provide food for the Roman Empire.

One of the world's longest lasting trade histories is the trade of dry cod from the Lofoten area to the southern parts of Europe, Italy, Spain and Portugal. The trade in cod, which started during the Viking period or before, has been going on for more than 1000 years and is still important.

Today, fisheries are estimated to provide 16 percent of the world population's protein, and that is considerably elevated in some developing nations and in regions that depend heavily on the sea.

According to the Food and Agriculture Organization (FAO) of the United Nations, total world capture fisheries production in 2000 was 86 million tons. The top producing countries were, in order, the People's Republic of China (excluding Hong Kong and Taiwan), Peru, Japan, the United States, Chile, Indonesia, Russia, India, Thailand, Norway and Iceland. Those countries accounted for more than half of the world's production; China alone accounted for a third of the world's production. Of that production, over 90 percent was marine and less than ten percent was inland.

There are large and important fisheries worldwide for various species of fish, mollusks and crustaceans. However, a very small number of species support the majority of the world's fisheries. Some of these species are herring, cod, anchovy, tuna, flounder, mullet, squid, shrimp, salmon, crab, lobster, oyster and scallops. All except these last four provided a worldwide catch of well over a million metric tons in 1999, with herring and sardines together providing a catch of over 22 million metric tons in 1999. Many other species as well are fished in smaller numbers.

For communities, fisheries provide not only a source of food and work but also a community and cultural identity.

Methods

A fishery can consist of one man with a small boat hand-casting nets, to a huge fleet of trawlers processing tons of fish per day. Some techniques are trawling, seining, driftnetting, handlining, longlining, gillnetting, dragger, tile, and diving.

Important Issues and Topics in Fisheries

There are many environmental issues surrounding fishing. These can be classed into issues that involve the availability of fish to be caught, such as overfishing, sustainable fisheries, and fisheries management; and issues surrounding the impact of fishing on the environment, such as by-catch (accidental catching of aquatic organisms other than the targeted species or juvenile members of the intended species).

There is an apparent and growing disparity between the availability of fish to be caught and humanity's desire to catch them, a problem that is exacerbated by the rapidly growing world population. As with some other environmental issues, often the people engaged in the activity of fishing—the fishermen—and the scientists who study fisheries science, who are often acting as fishery managers, are in conflict with each other, as the dictates of economics mean that fishers have to keep fishing for their livelihood, but the dictates of sustainable science mean that some fisheries must close or be reduced to protect the health of the population of the fish themselves. It is starting to be realized, however, that these two camps must work together to ensure fishery health through the twenty-first century and beyond.

The status of the world's fishery resources remains very much uncertain. The cover story of the May 15, 2003 issue of the science journal Nature—with Dr. Ransom A. Myers, an internationally prominent fisheries biologist (Dalhousie University, Halifax, Canada) as the lead author—was devoted to a summary of the scientific information. Myers and Worm assert that "industrialized fisheries typically reduced community biomass by 80 percent during the first 15 years of exploitation" and that "large predatory fish biomass today is only about ten percent of pre-industry levels." These large ocean fish are species at the top of the food chains (e.g., tuna, cod, among others). However, this article was subsequently criticized as being fundamentally flawed, at least in the case of tuna fisheries, and in particular tropical tuna fisheries.

In mid-October 2006, U.S. President George W. Bush joined other world leaders calling for a moratorium on deep-sea trawling, a practice shown to often have harmful effects on sea habitat, hence on fish populations.

The journal Science published a four-year study in November 2006, in which the authors conclude that loss of marine biodiversity is increasingly impairing the capacity of the coeans to provide food, recover from perturbations, and maintain water quality. The authors predict that, at prevailing trends, the world would run out of wild-caught seafood in 2048. The scientists stated that the decline was a result of overfishing, pollution and other environmental factors that were reducing the population of fisheries at the same time as their ecosystems were being degraded. Yet again the analysis has met criticism as being fundamentally flawed, and many fishery management officials, industry representatives and scientists challenge the findings. Siebert et al. conclude that the current biomass ranges from 36 percent to 91 percent of the biomass predicted in the absence of fishing, which is consistent or higher than standard fisheries management targets.

The debate continues. Even Worm et al. predict that the trends of accelerating declines are reversible, and note that many countries, such as Tonga, the United States and New Zealand, and international management bodies have taken steps to appropriately manage marine resources.

ARTISANAL FISHERY

Artisanal fisheries are the traditional fisheries involving fishing households (as opposed to commercial companies), using relatively small amount of capital and energy, relatively small fishing vessels (if any), making short fishing trips, close to shore, mainly for local consumption. In practice, definition varies between countries, e.g. from gleaning or a one-man canoe in poor developing countries, to more than 20-m. trawlers, seiners, or long-liners in developed ones. Artisanal fisheries can be subsistence or commercial fisheries, providing for local consumption or export. They are sometimes referred to as small-scale fisheries.

The line between what is artisanal and what is industrial can be quite blurry at times so it is best to look at it as a sliding scale and not just black and white. By comparing the size of the boats used with the amount of capital invested per man-on board it is easier to see these divisions. Meaning that even a large wooden canoe, that is man powered and using handlines would qualify as artisanal due to the small technological investment. While a small ultra-light boat equipped with the latest global positioning systems, downriggers, and sonar could still qualify as industrial.

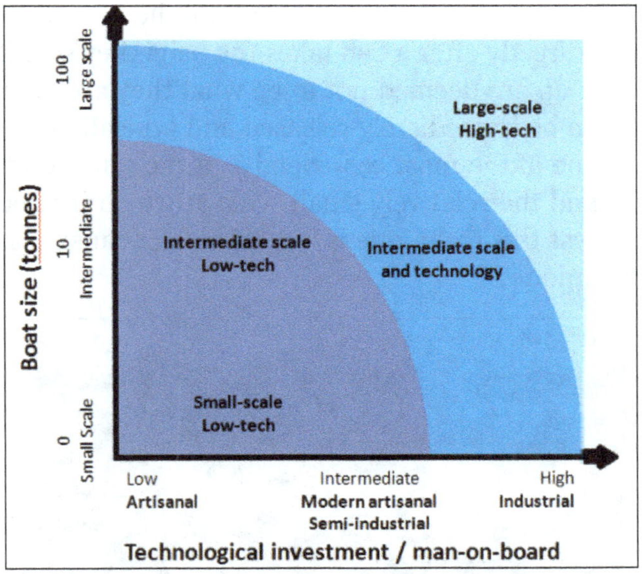

Modified and enlarged from the FAO artisanal fishery glossery.

Another key feature differentiating industrial and artisanal fishing is their purpose in fishing. Typically the goal of an industrial fishery is to catch as many fish as possible for economic gain. This carries with it various risks so as overfishing of a fish population causing the fishery to crash, among several other dangers. While it is true that artisanal fishers can sell a portion of their catch at markets, typically they primarily subsistence fish. Meaning they go fishing to catch enough food to feed to their families and others in their area. This smaller scale, more ecologically friendly goal, carries with it less of a risk of endangering fish populations, though that is not to say that all artisanal fisheries are "better" then all industrial fisheries. There are several artisanal fisheries that do pose a great risk to several ecosystems and marine species. Often it is what technology is used, how it is employed, and how the fishery is managed that defines how eco-friendly an artisanal fishery is.

Techniques used in Artisanal Fisheries

For the most part artisanal fisheries can be as varied as commercial fisheries and they often use similar techniques for the capture of their target species. As such, in order to better understand artisanal fisheries you must understand how most fisheries work in general. Below is a list of some of the most common fishing techniques used in artisanal fishing. Many of the issues described are much less severe in artisanal fishing due to their smaller scale but they still may be present.

Hook and Line

The technique that typically immediately comes to mind when one thinks of fishing. Using this technique a single person uses a relatively small numbers of lines and hooks to catch a fish. This ranges from a single line with a hook on the end which is brought up by hand, to the using a rod for the same purpose, all the way to the more advanced systems used for trolling such as downriggers. Though the technology varies greatly they all work on the same principle. Basically a single fisher is in direct control of a small number of lines each containing only a few hooks. These hooks can be baited or have lures designed to catch only specific kinds of fish, helping to reduce bycatch. Hooks can vary in size so that by using large enough hooks the fishermen can avoid catching juveniles of their target species, which in and of itself can help maintain the population's numbers. Since each line is manually brought up directly after a fish takes the baits they spend relatively little time on the line and if when brought up are deemed not to be what they are after, whether it be bycatch, improper size or sex they can be immediately released and typically have a reasonable chance of survival. This technique is one of the most eco-friendly in the sense that it has less of an impact largely due to the features and the relatively small scale at which it affects a population. That is not to say that if used in excess this technique is not capable of causing as much stress of the fish population as any other technique.

Use of a hook and a line with various deploying techniques.

Purse Seining

Purse seining is used to capture large schools of pelagic fish by encircling them with a large net. A small boat pulls the net off of the larger stationary one, and surrounds the school before coming back to the large ship. Along the bottom, deepest part, of the net runs a line that can be pulled tight causing the bottom of the net to close much like a purse, which traps all fish inside from escaping. The larger vessel can then draw in the net and take the captured fish on-board. Some smaller purse seins actually take the entire catch to shore while the fish remain in the sein. This is especially useful in remote areas without a port.

Depiction of a closed purse sein with its catch.

A downside to this technique is that it catches everything inside the purse, including some non-target species that feed on these schools such as billfish and dolphins. Modern practices have reduced this added risk by lowering parts of the net and allowing the non-target species to escape.

Gill Nets

Fish gills trapped in net allows for capture.

As their name implies, gill nets capture fish by taking advantage of a fish's gills. Basically, as a fish is swimming around it encounters the net. The mesh size of a gill net is completely uniform which makes it very selective for certain sized fish. If the fish is to small it simply goes through the mesh without any consequence, while if a fish is too large it cannot fit its head through the mesh deep enough to reach its gills so it is not trapped. But if a fish small enough so that it can fit its head through deep enough to reach its gills but too large for the rest of its body to fit through it become stuck. The gills of a fish are angled backwards so that if the fishes tries to swim backwards to free itself from the net the mesh enters the slit, where water exits, between the outer portion of the gill (the operculum) and the rest of the body effectively capturing the fish until the fisher pulls it out. Since the mesh can be size specific it can avoid capturing juvenile fish which have yet to reproduce and it avoids capturing larger fish that can be important in keeping the fish population plentiful since larger fish have more reproductive success. Gill nets, especially the now illegal drifting gill net, are the cause of some of the most severe seabird bycatch, but in some fisheries they still pose a serious risk for marine mammals, such as dolphins.

Longline

Longlining can best be described as a massive scale hook and line operation. Here a single long line is let off from a boat and periodically another line runs off of this main line where one to several baited hooks or lures are attached. This process is continued for very long distances, upwards of 40 miles (though typically much smaller for small-scale artisanal fishing). This technique is relatively inexpensive and not very labor intensive as machines are used for hauling the line in and out and large tracts of water can be covered fairly easily.

But the issue of bycatch is considerably more prevalent due to the larger scale. A huge issue is catching sea turtles, marine mammals, and seabirds on the hooks, where in the time that it takes to be brought to the surface they can suffocate. Circle hooks have reduced this issue, especially for sea turtles but it is still a problem, while hanging streamers from lines reduce seabird mortality.

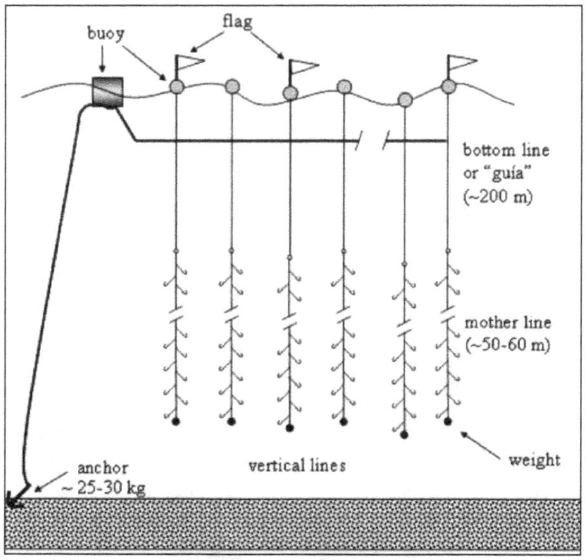

Modified after Moreno et al.

Dredging

Dragged behind a ship for bottom species.

This is one of the most common methods used to obtaining organisms on or near the ocean bottom. In its simplest form a dredge is a bag shaped net with a weighted front end that acts as a plow. The front edge digs into the ocean bottom and scoops everything in its path into the basket where it can later be hauled up and target organisms can be sorted out. The main targets in this type of fishing is crustaceans, like crabs and shrimp, and mollusks, mainly bivalves such as mussels, oysters, and clams. The techniques is very effective at obtaining these organisms but the problem lies in the damage it causes to the environment in obtaining these organisms.

The dredge is indiscriminate in what it picks up, it just continues plowing through everything in this path. This destroys or displaces countless other species in the area. This is of special concern where corals are present, since a dredge can destroy the coral and it can take many years for the area to recover the important habitat for other organisms. Also, the bivalves that are removed are very important for other organisms; whether they act as a substrate for them to attach, places to hide, or as a nursery ground for several organisms including salmon. The issue is that though this technique is destructive the only other option is diving which can be used in some areas quite effectively while in other it would be impractical to impossible.

Traps and Pots

Traditional trap used to catch target species.

This technique is far more common and varied in artisanal fishing then it is in industrial fishers. Other then species such as crab, lobster, and a few fish species traps are rarely used in these larger scale operations. But for an artisanal fisher it provides a relatively easy way to catch the target species. Though the design of a trap varies greatly depending on the location and what is being caught the design is basically the same. The organism swims through an opening whether due to just passive swimming or attracted to some sort of food, and is unable to escape. These traps tend to be designed to target specific species, due to size, bait, or position in the ecosystem and often exclude juvenile fish either through the mechanism of the trap or due to escape holes designed to let smaller fish out. The fish remain in the trap until the fisherman returns to check his trap. This means that if a non-target species was captured other than an increase in stress during the capture it can be released with high likelihood of being perfectly healthy. Due to how the traps work and their small scale they are one of the cleanest techniques available but are limited in the extent of their use.

Fishing from Land

This is quite possibly the most simple form of fishing. Here the fisher man remains of land and employs various techniques to capture fish. This ranges all the way from spearfishing, to using a bow and arrow, to hook and line fishing from shore, to throwing small nets, all the way up to using horses to drag a net along the shallow water near a beach. These techniques are very limited as they can only target species that are in rivers or near shore. They are also quite eco-friendly because in most of the techniques the fisher actually see exactly what he is catching before even attempting the capture. Due to many of these attributes shore fishing is almost solely used for subsistence feeding of the fisherman's family and those close by.

Hawaiian spearfishing.

Dynamite and Cyanide

These are some of the most outright destructive practices in artisanal fishing. The issue it that the techniques are amazingly effective at catching massive numbers of fish very cheaply, which is very important for a poor artisanal fisherman who worries more about feeding his family in the here and now and not the damage he is causing for the future. Dynamite is most dangerous when used on coral reefs, for although the blast kills all the fish in the area (or in some cases stuns the fish for the aquarium trade), allowing for massive quick profit, it also destroys the coral below. By destroying this slow growing habitat that the target species need to survive, you effectively remove the area of your target species in only a few catches.

Cyanide poisoning is used mainly in the live fish fisheries and capturing fish for the aquarium trade. The chemical sodium cyanide is injected into a target species den which temporarily stuns the fish allowing it to be easily captured. The issue is that the cyanide affects other non-target species and the coral itself, often killing them. Continued used of cyanide can be just as devastating as any other form of fishing, including dynamite

General Comparison to Industrial Fishing

The figure depicts that overall artisanal fisheries catch as much fish that are actually eaten as food as the industrial company even though they are so much smaller and have less money invested in them. Not only that but artisanal fisheries employ about 24 times more people, most in developing countries which may help reduce poverty. Also, artisanal fisheries use less than 1/7th of the fuel, even though they catch the same amount of food. The "annual catch reduced to fishmeal and oils" may need an explanation. Some fisheries target "trash" fish that typically are not used for human consumption. Instead they catch massive amounts of these fish and process them into fish food. Typically this fish food is used in aquaculture to feed fish that we like to eat, such as salmon and tuna. Aquaculture of these species was originally set up to alleviate some pressure from wild stocks, but as a result it is putting a new massive pressure on "trash" fish stocks, and since artisanal fisheries largely avoid this aspect entirely they also avoid all the issues associated with it. Lastly, due to the various practices used by artisanal fisheries, they acquire hardly any bycatch when compared to the massive numbers of non-target fish, seabirds, marine mammals, seaturtles, etc. that are killed by industrial fisheries.

FISHERY / BENEFITS	LARGE SCALE	SMALL SCALE
Subsidies	25-27 billion	5-7 billion
Number of fishers employed	about 1/2 million	over 12 million
Annual catch for human consumption	about 30 million t	same: about 30 million t
Annual catch reduced to fishmeal and oils	35 million t	Almost none
Annual fuel oil consumption	about 37 million t	about 5 million t
Catch per tonne of fuel consumed	1-2 t	4–8 t
Fish and other sealife discarded at sea	8–20 million tonnes	Very little

Modified after Jacquet et al.

Management of Artisanal Fisheries

Management of artisanal fisheries is difficult to say the least. Many artisanal fisheries are managed at a community level by laws or common practices that have been established and are enforced based on the interests of the people. For more information how community based management works and how some of these beliefs and practices are being used to improve more developed fisheries see: community fisheries management. Many artisanal fisheries feel that federal regulations of their fisheries are unnecessary and that these regulations would just hamper a process that is working perfectly fine on its own. For many artisanal fisheries the fisherman feel a sense of stewardship for their area of the sea. It is not just a way to make money, it is their very livelihood. As such, they often fell that they must protect it and undertake practices to best get what they need with as little negative impact as possible. But federal agencies have the issue of accounting for artisanal fisheries in their industrial fishery plans. As figure shows quite clearly, artisanal fisheries make up about half of the total amount of fish caught for food. So if fishery management groups did not take artisanal fisheries into account there would likely be disastrous consequences for the fish population stock.

That is not to say that all artisanal fisheries are perfect, far from it. Typically long standing subsistence artisanal fisheries who only fish to feed their families and neighbors pose little risk for most fish populations. But some fisheries have shifted to massive catches of target species for export to developed countries. An example of this would be the Mexican shark fishery that due to increase fishing pressure is seeing a shark population crash. Many groups are calling for immediate management of fisheries like this in fear of what might happen if they continue to go unregulated. But one of the largest issues with artisanal fisheries is due to how remote many of these fisheries can be. This remoteness not only makes it hard to get a clear idea of actual catch numbers, needed for proper fisheries management, but it makes it impractical if not impossible to enforce any of the management regulations if and when they are created.

A relatively new research network has been established with a primary goal being to conduct research into many aspects of artisanal fisheries in order to more effectively manage them.

Issue of Bycatch in Artisanal Fisheries

Non-target fish caught in a shrimp fishery.

If one had two pick the two biggest issues concerning most fisheries, artisanal and industrial, it would probably be overfishing and bycatch. While most management agencies try to account for both, typically more attention is paid to prevent overfishing as can be seen in the management section above. The easiest way to describe bycatch would be the capture of any species that the fisher is not specifically targeting and is typically discarded. Bycatch is accounted for in some fisheries, such as once a certain poundage of bycatch is reached that specific boat or in some cases the entire fishery is shut down. Plus, bycatch acts as an added cost due to the time and energy needed to remove the non-target species. With that prospective in mind, fishers typically try to reduce their bycatch by as much as possible. They do so by taking advantage of new fishing techniques and technology that has been shown to decrease their catch to bycatch ratios even if sometimes their catch amount might actually be less. This is because some types of technology allow non-target species to escape but as a result some of their target species may likewise escape. Consumers, however, are typically willing to pay more for fish caught using these more sustainable techniques so it may even out it the end.

The issue of bycatch is found in all fisheries but it is often less of a concern in artisanal fisheries. This is largely due to the scale of artisanal fisheries. There is less fishing pressure, often using species specific techniques, in defined areas typically only hours from port. Artisanal fisheries simply undergo more trips using safer techniques then their commercial counterparts who take massive trips using techniques designed for large catches at the expense of bycatch. Also, in many artisanal fisheries they feel a sense of stewardship of their sea and so enact practices to reduce their bycatch in order to protect their livelihood. Figure shows this idea quite clearly by the fact that even though artisanal fisheries catch just as much fish as industrial fisheries, due to these differences there is substantially less bycatch.

BASED ON TYPE OF WATER

Fresh Water Fisheries

Freshwater fish are those that spend some or all of their lives in fresh water, such as rivers and lakes, with a salinity of less than 0.05%.

Some fisheries such as the Great Lakes in North America and Lake Victoria in Africa support a commercial harvest that is exported and marketed away from the source. Other fisheries produce large quantities of fish, but most or all of the catch is consumed locally, so the economic importance is harder to measure. Examples of the latter include the Amazon and its tributaries in South America, the Chang Jiang (Yangtze River) in China, the Tonle Sap River in Cambodia, and the Mekong River in Vietnam.

The Great Lakes

The Great Lakes range from cold, deep, clear water with heavily forested shorelines such as Lake Superior to the warmer, shallower water of Lake Erie surrounded by farms and urban areas. The cold, clear waters produce desirable species such as whitefish, trout, and salmon. The warm waters of Lake Erie with their heavy nutrient load are more biologically productive, but the varieties of fish taken generally are less commercially desirable species, such as varieties of buffalo shad and white perch. Although Lake Erie still supports one of the largest commercial freshwater fisheries in the world, commercial fishing on all the Great Lakes is declining.

Lake Victoria

Africa's Lake Victoria is the world's largest tropical lake and the world's second largest fresh-water lake. In the past, Lake Victoria supported a wide variety of fish species as well as an important subsistence fishery along its shores. However, years of abuse have dramatically altered the lake. The shores have been cleared for cash crops of tea and coffee, while its marshes have been drained. Agricultural chemicals applied to the area around the lake eventually end up in the lake. The biggest change has come as the result of the introduction of the Nile perch. Frozen Nile perch fillets are sold throughout Europe and are a valuable export commodity. However, local small-scale fishing is being pushed aside by large-scale commercial fishing using modern technology.

The Amazon Basin

The Amazon is the longest river in the world. The Amazon River Basin supports more than 3,000 fish species—probably the greatest diversity of fresh-water fish of any single river basin. Fish provide 80 percent of the protein consumed by people in the Amazon region. Because of the high protein content of their diet, inhabitants along the river are much less likely to be malnourished than rural people in regions without fisheries. Fish from the Amazon also are an important export to Asian countries, especially Japan.

Steelhead trout are one of the important cold-water fish species in Lake Michigan. Lake Michigan steelheads are wild hybrids.

Chang Jiang

The Chang Jiang or "Great River" at 6,380 kilometers (3,960 miles) is the longest river in China and Asia, and the third longest in the world. It also supports a highly productive fishery. In recent years, the average annual catch has dropped from over 400,000 metric tons to about 100,000 metric tons.

Tonle Sap

The Tonle Sap is a wide lake in Cambodia. Tonle Sap means "Great Lake." Tonle Sap is also the name of the river that runs into the lake. It is the largest body of fresh water in Southeast Asia and also is one of the world's most productive inland fisheries. During floods, both the Tonle Sap and Mekong River water levels rise until the Mekong spills over into the Tonle Sap. This complex relation, along with the surrounding wet-lands , accounts for the extraordinary productivity of the lake–river complex. Most of the annual catch of 110,000 metric tons is consumed locally.

Mekong River

Most of the 60 million people who live in the Mekong Basin in Vietnam depend on fish for their entire animal protein intake. These local fisheries are vitally important for nutrition of some of the world's poorest people who otherwise would be unable to afford a high-protein diet. Sale of fish and fish products also is an important income supplement for many families living in the basin. By some estimates, the Mekong and its tributaries alone contribute 1 percent of the global fish harvest.

Threats to Fresh-water Fisheries

Worldwide, all fisheries are threatened by various factors. Pollution , loss of habitat , invasive species, parasites, illegal fishing methods, and overfishing are some of the more serious threats.

Pollution

All of the world's fresh-water fisheries face some type of threat from pollution. Lake Victoria is threatened by agricultural runoff and industrial waste. The Chang Jiang is the cleanest of China's rivers, but still threatened by pollution. And completion of the Three Gorges Dam (scheduled for 2009) will reduce flow in the lower river, which in turn may allow contaminated sediments to build up. In recent years, fishing on the Chang Jiang has been severely affected by environmental pollution, construction of irrigation projects, and the reclamation of land from lakes.

In North America, chemical pollution is a great concern in the Great Lakes region because of its heavy industrial and urban centers within larger agricultural areas. The Lake Erie catch of white-fish (a desirable species) collapsed in the 1960s, partly due to the effects of agricultural chemical runoff. Whitefish populations in Lake Superior and Lake Michigan already were reduced as a result of parasitism by sea lampreys. Since the 1960s, the white-fish catch has been steadily increasing as efforts to reduce water pollution and control the sea lamprey had been successful. At the same time as the whitefish take was declining, the tons of carp harvested per year steadily increased. This is one of the ironic aspects of this type of pollution. The lake remained very productive, but carp and the other kinds of fish taken were less economically desirable, so the income from commercial fishing dropped.

Habitat Alteration

Habitat alteration is possibly the most serious threat faced by fresh-water fisheries. For example, dams are being built in the Chang Jiang basin. Lake and river shores are being deforested in the Amazon River Basin, around Lake Victoria, and on the banks of the Tonle Sap. Lakes are being filled to create new agricultural land on the Chang Jiang. All of these activities degrade the quality of the fishery. However, managing these problems is a complex issue involving socioeconomic and political factors as well as fisheries management strategies.

Invasive Species and Parasites

More than 140 invasive species have been identified in the Great Lakes. Most have appeared since the St. Lawrence Seaway opened in 1959. Sea lampreys in the Great Lakes are a notable example of the damage that can be done by an invasive species, and of the success of fisheries management strategies.

Because sea lampreys are not native to the Great Lakes, there were no natural controls and their numbers rose dramatically. Parasitism by sea lampreys nearly destroyed the whitefish and lake trout fisheries by the 1950s. Control efforts eventually resulted in a 90-percent reduction of sea lamprey populations in most areas, generally restoring a healthy environment for fish survival and spawning. However, ongoing control costs the United States and Canada more than $12 million annually.

Controlling the sea lamprey is a success story, but other invasive species such as the zebra and quagga mussels remain serious problems. Controlling invasive species and preventing future invasions must remain a top priority for fisheries managers.

Fisheries biologists use a weak electric current to stun fish temporarily so they can be identified, measured, and tagged. Accurately assessing fish populations is a critical component of fisheries management.

Parasites are of concern for several reasons. Parasites may have been keeping a nuisance species in check in its native habitat. In the absence of those parasites and other predators, the nuisance species can grow exponentially. On the other hand, fish weakened by nonnative parasites, such as the sea lamprey, are likely to die either from their weakened condition or from secondary infection.

Exotic species also can introduce nonnative parasites into an ecosystem. An example of this is whirling disease, a potentially fatal condition that appears in young trout and salmon. The disease, which may have been introduced through fish stocking programs, is caused by a nonnative parasite that infects the fish cartilage and nervous system. It is found in at least eleven western U.S. states. The Madison River in Montana has suffered a 90-percent decline in rainbow trout populations since the disease first appeared. Several parasites of mussels also have been accidentally introduced into U.S. fresh waters.

Illegal Fishing Methods

Among other threats, the Chang Jiang River ecosystem is threatened by illegal fishing activities. In 1999, about 3,000 people were arrested for using poisons, electrical shocking systems, and other illegal methods to take fish. The fishers are trapped in an economic spiral. Pollution, dams, and draining of lakes, all ecosystem threats in themselves, reduce fish populations. This forces the fishers to turn to more economical and possibly illegal methods to catch more fish. Other fisheries worldwide are threatened by illegal fishing.

Overfishing and Depleted Wild Stocks

A notable example of the consequences of overfishing is the herring fishery in the Great Lakes, which has been severely depleted. Similarly, the abundance and variety of Lake Victoria fish has dramatically decreased due to overfishing. In the Tonle Sap Lake and River, competition between government licensed commercial fishermen and local people who fish for food for their families (i.e., subsistence fishing) has led to overfishing.

Fresh Water Fisheries Management

Modern fresh water fisheries management takes an ecosystem approach. This approach to fisheries management recognizes the interrelationships between air, land, water, and all living organisms, rather than focusing on a single aspect. Ecosystem-based fisheries management tries to move beyond political boundaries and look at natural boundaries, such as watersheds (drainage basins), as the unit of management. An ecosystem approach includes the whole system, not just parts of it. Ecosystem approaches also incorporate the concept of sustainability . Ecosystem approaches look for long-term solutions rather than short-term fixes.

Strategies that can be used successfully by ecosystem-based fisheries management include temporary fishing bans, catch limits, size limits, and habitat improvement. In addition, captive breeding and restocking may be used to rebuild depleted populations. Strict enforcement of laws and catch limits also may help. For example, the countries surrounding Lake Victoria have instituted a series of temporary fishing bans to allow wild stocks (populations) time to recover. But agricultural runoff and industrial pollution continue to threaten Lake Victoria. As of 2002, Lake Victoria remained a severely damaged ecosystem that will require dramatic efforts to rescue.

The ecosystem management approach has been most successful in the Great Lakes. While serious problems remain, there are several successes. The herring stocks are rebounding due to fishing bans and restocking programs. Various kinds of water pollution have been reduced or eliminated.

The Future of Fresh Water Fisheries

Commercial fresh water fisheries have never been as economically important as commercial marine fisheries, except in certain local areas such as the Mekong River Basin. Only about one-fifth of the world's total fish catch comes from fresh-water fisheries. Furthermore, the economic importance of fresh-water fisheries worldwide is diminishing. Depleted fish populations and various other threats continue to make the harvesting of wild stocks uneconomical. Aquaculture is becoming increasingly important and is displacing the commercial harvesting of wild stocks. Capture fisheries now account for less than half of all fresh-water fish consumed globally.

Marine Fisheries

Around half of the world's fish species are found in the marine environment, and those that are edible have long provided an important food resource for humans and other animals. Fishing is now a huge commercial operation and marine fisheries have grown in size and sophistication in the last 50 years. Fish are caught in nets or on lines, sorted on board a fishing vessel, and then brought back to shore.

Although the oceans are vast, their fish stocks are not limitless. Many marine fisheries are now taking fish faster than stocks can renew themselves. This means that some fishes, such as the Atlantic cod, are now regarded as endangered species and several marine ecosystems are under threat. Therefore, many nations have now put limits on fishing, although it is not known to what extent fish stocks can recover. Consumers can help by purchasing from sustainable marine fisheries, which catch only those species whose populations are not threatened.

There is a long tradition of fishing, both for pleasure and for food, around the world. Fish are highly valued as a source of animal protein, comparing favorably with meat in terms of cost and health benefits. Many health authorities actively promote regular consumption of fish, such as salmon and sardines, that are rich in heart-healthy omega-3 fatty acids. The average consumption of fish worldwide is about 30 pounds a week. Fishing covers harvesting fish themselves, shellfish, and other aquatic animals, such as squid and octopus, but not whales. Some marine fish are farmed in tanks, rather than in the open water. Besides food, fish are used for their oils, enzymes, and other manufactured products, as well as having a decorative and educational function in aquaria.

Of the 30,000 or so of identified fish species, about 500 are known to be edible. Around half of all known fish species live in a marine environment; the rest are found in freshwater. Tuna is found in the open ocean, whitefish and salmon around temperate coastlines, shrimp in coastal seas and oysters, clams, and mussels in estuaries. There are also areas called upwellings, where nutrient-rich deep water comes to the surface and can support high fish productivity. These are found off the western coasts of Africa, North and South America, across the Western Pacific, and around the Southern Ocean and are a popular site for marine fisheries. Countries with major fish industries include China, the United States, Chile, Iceland, and Norway.

Because the seas and oceans are so large, covering 70% of Earth's surface, it was once assumed that the supply of marine organisms to meet human needs is limitless. The fishing industry therefore expanded dramatically from the 1960s, with more vessels being put to sea and advances in fishing technology bringing in higher yields of fish. The way fish are caught varies with the size of the operation and the type of fish involved. Most fishing involves pots, lines, and nets, but the size of the vessel may vary from a single fisher with a few lines to massive ocean-going trawlers casting thousands of square miles of strong plastic netting each day. Fish harvesting has increased 5 fold in the last 50 years. From 1970 to 1990, it went up from 60 million to 80 million tons per year and then leveled off. Fish now account for about 16% of animal protein consumption by humans, and the industry employs about 10 million people directly, with many millions more being dependent on it for sustenance.

Removal of large numbers of fish species cause shifts in the marine food web. There are other effects on the marine ecosystem. Fishers divide their catch into two groups. Undersized and diseased fish that will not sell are thrown back to avoid taking up valuable space on the fishing vessel. The keepers, as they are known, are taken back to shore. The discards provide a food source for large sea birds, like the great skua in the North Sea. However, fewer of these discards have been available in recent years, because fish hauls have been restricted. Great skua populations have declined and those birds remaining have had to turn to alternative food sources, such as other, smaller, sea birds.

Impacts and Issues

The unsustainable approach of marine fisheries over the last 50 years or so has now hit fish stocks hard. The marine environment can support the taking of around 100 million tons of fish a year. The global fishing fleet is now about two and a half times larger than what the oceans can sustainably support. Motors have replaced sails, boats are bigger, and onboard refrigerators allow fishers to stay at sea longer.

Over 1,000 species of fish are now endangered, either directly because of overfishing or because of the impact that marine fisheries have upon ecosystems. One of the most important, economically, is the Atlantic cod. Large open ocean fish, like tuna, swordfish, and marlin, are also in trouble, along with large groundfish such as halibut, skate, and flounder. Stocks of all these are down to around 10% according to the FAO.

Meanwhile, fish can even pose a health hazard. Fish species exposed to contaminated waters may concentrate toxins in their flesh, including pesticides, mercury, and polychlorinated biphenyls. These substances may enter the water at low concentrations, but as they are absorbed through the marine food chain, they accumulate to levels that may be harmful. Farmed fish may also be contaminated with the antibiotics used to ensure high yields.

Marine fisheries can give rise to other problems. Sometimes they harm or destroy coastal habitats, mangrove swamps, and coral reefs, the latter being especially endangered by bottom trawling. The method by which fish are harvested is also not always sustainable. For instance, nets used to catch shrimp, tuna, and scallops may harm other species. Unusual fish species from coral reefs may be taken to stock aquaria. Meanwhile, fish farms may deplete wild fish stocks when they take supplies for ponds. There are, however, a few fish species left that can be taken sustainably from the marine environment. These include farmed catfish or tilapia, Pacific pollack, squid, crayfish, crabs, and wild salmon.

The way forward for marine fisheries is to become more sustainable and consumers more aware. The Environmental Defense Fund recommends avoiding top predator species including swordfish, marlin, shark, bluefin, and albacore tuna when choosing a seafood meal. These species grow slowly and take several years to reproduce. Groundfish and deepwater fish that have been affected by overfishing include orange roughy; grouper; flounder or sole; cod; haddock; pollack, which is used in various fish products; and monkfish.

Many countries where marine fishing is an important industry now have legislation in place to protect stocks. In the United States, this is the Magnuson-Stevens Fishery Conservation and Management Reauthorization Act of 2006, which sets catch limits and requires sustainable management to protect against over-fishing.

BASED ON THE OBJECT

Aquaculture of Salmonids

The aquaculture of salmonids is the farming and harvesting of salmonids under controlled conditions for both commercial and recreational purposes. Salmonids (particularly salmon and rainbow trout), along with carp, and tilapia are the three most important fish species in aquaculture. The most commonly commercially farmed salmonid is the Atlantic salmon. In the U.S. Chinook salmon and rainbow trout are the most commonly farmed salmonids for recreational and subsistence fishing through the National Fish Hatchery System. In Europe, brown trout are the most commonly reared fish for recreational restocking. Commonly farmed nonsalmonid fish groups include tilapia, catfish, sea bass, and bream.

In 2007, the aquaculture of salmonids was worth US$10.7 billion globally. Salmonid aquaculture production grew over ten-fold during the 25 years from 1982 to 2007. In 2012, the leading producers of salmonids were Norway, Chile, Scotland and Canada.

Much controversy exists about the ecological and health impacts of intensive salmonids aquaculture. Of particular concern are the impacts on wild salmon and other marine life. Some of this controversy is part of a major commercial competitive fight for market share and price between Alaska commercial salmonid fishermen and the rapidly evolving salmonid aquaculture industry.

Methods

Assynt salmon hatchery, near Inchnadamph in the Scottish Highlands.

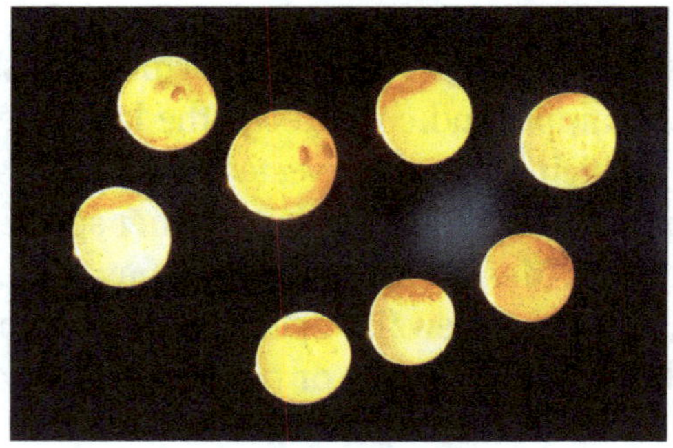

Very young fertilised salmon eggs, notice the developing eyes and vertebral column.

The aquaculture or farming of salmonids can be contrasted with capturing wild salmonids using commercial fishing techniques. However, the concept of "wild" salmon as used by the Alaska Seafood Marketing Institute includes stock enhancement fish produced in hatcheries that have historically been considered ocean ranching. The percentage of the Alaska salmon harvest resulting from ocean ranching depends upon the species of salmon and location. Methods of salmonid aquaculture originated in late 18th-century fertilization trials in Europe. In the late 19th century, salmon hatcheries were used in Europe and North America. From the late 1950s, enhancement programs based on hatcheries were established in the United States, Canada, Japan, and the USSR. The contemporary technique using floating sea cages originated in Norway in the late 1960s.

Salmon egg hatching: In about 24 hr, it will be a fry without the yolk sac.

Salmonids are usually farmed in two stages and in some places maybe more. First, the salmon are hatched from eggs and raised on land in freshwater tanks. Increasing the accumulated thermal units of water during incubation reduces time to hatching. When they are 12 to 18 months old, the smolt (juvenile salmon) are transferred to floating sea cages or net pens anchored in sheltered bays or fjords along a coast. This farming in a marine environment is known as mariculture. There they are fed pelleted feed for another 12 to 24 months, when they are harvested.

Norway produces 33% of the world's farmed salmonids, and Chile produces 31%. The coastlines of these countries have suitable water temperatures and many areas well protected from storms. Chile is close to large forage fisheries which supply fish meal for salmon aquaculture. Scotland and Canada are also significant producers; and it was reported in 2012 that the Norwegian government at that time controlled a significant fraction of the Canadian industry.

Modern salmonid farming systems are intensive. Their ownership is often under the control of huge agribusiness corporations, operating mechanized assembly lines on an industrial scale. In 2003, nearly half of the world's farmed salmon was produced by just five companies.

Hatcheries

Modern commercial hatcheries for supplying salmon smolts to aquaculture net pens have been shifting to recirculating aquaculture systems (RAS)s where the water is recycled within the hatchery. This allows location of the hatchery to be independent of a significant fresh water supply and allows economical temperature control to both speed up and slow down the growth rate to match the needs of the net pens.

Conventional hatchery systems operate flow-through, where spring water or other water sources flow into the hatchery. The eggs are then hatched in trays and the salmon smolts are produced in raceways. The waste products from the growing salmon fry and the feed are usually discharged into the local river. Conventional flow-through hatcheries, for example the majority of Alaska's enhancement hatcheries, use more than 100 tonnes (16,000 st) of water to produce a kg of smolts.

An alternative method to hatching in freshwater tanks is to use spawning channels. These are artificial streams, usually parallel to an existing stream with concrete or rip-rap sides and gravel bottoms. Water from the adjacent stream is piped into the top of the channel, sometimes via a header pond to settle out sediment. Spawning success is often much better in channels than in adjacent streams due to the control of floods which in some years can wash out the natural redds. Because of the lack of floods, spawning channels must sometimes be cleaned out to remove accumulated sediment. The same floods which destroy natural redds also clean them out. Spawning channels preserve the natural selection of natural streams as no temptation exists, as in hatcheries, to use prophylactic chemicals to control diseases. However, exposing fish to wild parasites and pathogens using uncontrolled water supplies, combined with the high cost of spawning channels, makes this technology unsuitable for salmon aquaculture businesses. This type of technology is only useful for stock enhancement programs.

Sea Cages

Sea cages, also called sea pens or net pens, are usually made of mesh framed with steel or plastic. They can be square or circular, 10 to 32 m (33 to 105 ft) across and 10 m (33 ft) deep, with volumes between 1,000 and 10,000 m^3 (35,000 and 353,000 cu ft). A large sea cage can contain up to 90,000 fish.

They are usually placed side by side to form a system called a seafarm or seasite, with a floating wharf and walkways along the net boundaries. Additional nets can also surround the seafarm to keep out predatory marine mammals. Stocking densities range from 8 to 18 kg (18 to 40 lb)/m^3 for Atlantic salmon and 5 to 10 kilograms (11 to 22 lb)/m^3 for Chinook salmon.

In contrast to closed or recirculating systems, the open net cages of salmonid farming lower production costs, but provide no effective barrier to the discharge of wastes, parasites, and disease into the surrounding coastal waters. Farmed salmon in open net cages can escape into wild habitats, for example, during storms.

An emerging wave in aquaculture is applying the same farming methods used for salmonids to other carnivorous finfish species, such as cod, bluefin tuna, halibut, and snapper. However, this is likely to have the same environmental drawbacks as salmon farming.

A second emerging wave in aquaculture is the development of copper alloys as netting materials. Copper alloys have become important netting materials because they are antimicrobial (i.e., they destroy bacteria, viruses, fungi, algae, and other microbes), so thery prevent biofouling (i.e., the undesirable accumulation, adhesion, and growth of microorganisms, plants, algae, tubeworms, barnacles, mollusks, and other organisms). By inhibiting microbial growth, copper alloy aquaculture cages avoid costly net changes that are necessary with other materials. The resistance of organism growth on copper alloy nets also provides a cleaner and healthier environment for farmed fish to grow and thrive.

Feeding

Salmonids are carnivorous and are currently being fed compound fish feeds containing fish meal and other feed ingredients, ranging from wheat byproducts to soybean meal and feather meal. Being aquatic carnivores, salmonids do not tolerate or properly metabolize many plant-based carbohydrates and use fats instead of carbohydrates as a primary energy source.

With the amount of worldwide fish meal production being almost a constant amount for the last 30+ years and at maximum sustainable yield, much of the fish meal market has shifted from chicken and pig feed to fish and shrimp feeds as aquaculture has grown in this time.

Work continues on developing salmonid diet made from concentrated plant protein. As of 2014, an enzymatic process can be used to lower the carbohydrate content of barley, making it a high-protein fish feed suitable for salmon. Many other substitutions for fish meal are known, and diets containing zero fish meal are possible. For example, a planned closed-containment salmon fish farm in Scotland uses ragworms, algae, and amino acids as feed. Some of the eicosapentaenoic acid and docosahexaenoic acid (in Omega-3 fatty acids) may be replaced by land-based (non-marine) algae oil, reducing the harvest of wild fish as fish meal.

However, commercial economic animal diets are determined by least-cost linear programming models that are effectively competing with similar models for chicken and pig feeds for the same feed ingredients, and these models show that fish meal is more useful in aquatic diets than in chicken diets, where they can make the chickens taste like fish. Unfortunately, this substitution can result in lower levels of the highly valued omega-3 content in the farmed product. However, when vegetable oil is used in the growing diet as an energy source and a different finishing diet containing high omega-3 content fatty acids from either fish oil, algae oils, or some vegetable oils are used a few months before harvest, this problem is eliminated.

As of 2008, 50-80% of the world fish oil production is fed to farmed salmonids.

Farm raised salmonids are also fed the carotenoids astaxanthin and canthaxanthin, so their flesh colour matches wild salmon, which also contain the same carotenoid pigments from their diet in the wild.

On a dry-dry basis, 2–4 kg of wild-caught fish are needed to produce 1 kg of salmon. The ratio may be reduced if non-fish sources are added. Wild salmon require about 10 kg of forage fish to produce 1 kg of salmon, as part of the normal trophic level energy transfer. The difference between the two numbers is related to farmed salmon feed containing other ingredients beyond fish meal and because farmed fish do not expend energy hunting.

In 2017 it was reported that the American company Cargill has been researching with EWOS on alternative feeds in its RAPID feed and COMPASS programs in Norway. These methods studied macronutrient profiles of fish feed based upon geography and season. Using RAPID feed, salmon farms reduced the time to maturity of salmon to about 15 months, in a period one-fifth faster than usual.

Harvesting

Modern harvesting methods are shifting towards using wet-well ships to transport live salmon to the processing plant. This allows the fish to be killed, bled, and filleted before rigor has occurred. This results in superior product quality to the customer, along with more humane processing. To obtain maximum quality, minimizing the level of stress is necessary in the live salmon until actually being electrically and percussively killed and the gills slit for bleeding. These improvements in processing time and freshness to the final customer are commercially significant and forcing the commercial wild fisheries to upgrade their processing to the benefit of all seafood consumers.

An older method of harvesting is to use a sweep net, which operates a bit like a purse seine net. The sweep net is a big net with weights along the bottom edge. It is stretched across the pen with the bottom edge extending to the bottom of the pen. Lines attached to the bottom corners are raised, herding some fish into the purse, where they are netted. Before killing, the fish are usually rendered unconscious in water saturated in carbon dioxide, although this practice is being phased out in some countries due to ethical and product quality concerns. More advanced systems use a percussive-stun harvest system that kills the fish instantly and humanely with a blow to the head from a pneumatic piston. They are then bled by cutting the gill arches and immediately immersing them in iced water. Harvesting and killing methods are designed to minimize scale loss, and avoid the fish releasing stress hormones, which negatively affect flesh quality.

Wild versus Farmed

Wild salmonids are captured from wild habitats using commercial fishing techniques. Most wild salmonids are caught in North American, Japanese, and Russian fisheries. The following table shows the changes in production of wild salmonids and farmed salmonids over a period of 25 years, as reported by the FAO. Russia, Japan and Alaska all operate major hatchery based stock enhancement programs that are really ocean ranching. The resulting fish hatchery fish are defined as "wild" for FAO and marketing purposes.

Salmonid production in tonnes by species					
	1982		2007		2013
Species	Wild	Farmed	Wild	Farmed	
Atlantic salmon	10,326	13,265	2,989	1,433,708	2,087,110
Steelhead		171,946		604,695	
Coho salmon	42,281	2,921	17,200	115,376	
Chinook salmon	25,147		8,906	11,542	
Pink salmon	170,373		495,986		
Chum salmon	182,561		303,205		
Sockeye salmon	128,176		164,222		

Total salmonid production				
	1982		2007	
	tonnes	percent	tonnes	percent
Wild	558,864	75	992,508	31
Farmed	188,132	25	2,165,321	69
Overall	746,996		3,157,831	

Issues

The US in their dietary guidelines for 2010 recommends eating 8 ounces per week of a variety of seafood and 12 ounces for lactating mothers, with no upper limits set and no restrictions on eating farmed or wild salmon. In 2018, Canadian dietary guidelines recommended eating at least two servings of fish each week and choosing fish such as char, herring, mackerel, salmon, sardines, and trout.

Currently, much controversy exists about the ecological and health impacts of intensive salmonid aquaculture. Of particular concern are the impacts on wild salmonids and other marine life and on the incomes of commercial salmonid fishermen. However, the 'enhanced' production of salmon juveniles – which for instance lead to a double-digit proportion (20-50%) of the Alaska's yearly 'wild' salmon harvest - is not void of controversy, and the Alaska salmon harvest are highly dependent on the operation of Alaska's Regional Aquaculture Associations. Furthermore, the sustainability of enhanced/hatchery-based 'wild' caught salmon has long been hotly debated, both from a scientific and political/marketing perspective. Such debate and positions were central to a 'halt' in the re-certification of Alaska salmon fisheries by the Marine Stewardship Council (MSC) in 2012. The Alaska salmon fisheries subsequently re-attained MSC-certification status; however the heavily hatchery-dependent Prince William Sound (PWS) unit of certification ("one of the most valuable fishing area in the State") was for several years excluded from the MSC-certification (it remained 'under assessment' pending further analysis).

Disease and Parasites

In 1972, *Gyrodactylus*, a monogenean parasite, was introduced with live trout and salmon from Sweden (Baltic stocks are resistant to it) into government-operated hatcheries in Norway. From the hatcheries, infected eggs, smolt, and fry were implanted in many rivers with the goal to strengthen the wild salmon stocks, but caused instead devastation to some of the wild salmon populations affected.

In 1984, infectious salmon anemia (ISAv) was discovered in Norway in an Atlantic salmon hatchery. Eighty percent of the fish in the outbreak died. ISAv, a viral disease, is now a major threat to the viability of Atlantic salmon farming. It is now the first of the diseases classified on List One of the European Commission's fish health regimen. Amongst other measures, this requires the total eradication of the entire fish stock should an outbreak of the disease be confirmed on any farm. ISAv seriously affects salmon farms in Chile, Norway, Scotland, and Canada, causing major economic losses to infected farms. As the name implies, it causes severe anemia of infected fish. Unlike mammals, the red blood cells of fish have DNA, and can become infected with viruses. The fish develop pale gills, and may swim close to the water surface, gulping for air. However, the disease can also develop without the fish showing any external signs of illness, the fish maintain a normal appetite, and then they suddenly die. The disease can progress slowly throughout an infected farm, and in the worst cases, death rates may approach 100%. It is also a threat to the dwindling stocks of wild salmon. Management strategies include developing a vaccine and improving genetic resistance to the disease.

In the wild, diseases and parasites are normally at low levels, and kept in check by natural predation on weakened individuals. In crowded net pens, they can become epidemics. Diseases and parasites also transfer from farmed to wild salmon populations. A recent study in British Columbia links the spread of parasitic sea lice from river salmon farms to wild pink salmon in the same river. The European Commission (2002) concluded, "The reduction of wild salmonid abundance is also linked to other factors but there is more and more scientific evidence establishing a direct link between the number of lice-infested wild fish and the presence of cages in the same estuary." It is reported that wild salmon on the west coast of Canada are being driven to extinction by sea lice from nearby salmon farms. These predictions have been disputed by other scientists and recent harvests have indicated that the predictions were in error. In 2011, Scottish salmon farming introduced the use of farmed wrasse for the purpose of cleaning farmed salmon of ectoparasites.

Globally, salmon production fell around 9% in 2015, in large part due to acute outbreaks of sea lice in Scotland and Norway. Lasers are used to reduce lice infections.

Pollution and Contaminants

Salmonid farms are typically sited in marine ecosystems with good water quality, high water exchange rates, current speeds fast enough to prevent pollution of the bottom but slow enough to prevent pen damage, protection from major storms, reasonable water depth, and a reasonable distance from major infrastructure such as ports, processing plants, and logistical facilities such as airports. Logistical considerations are significant, and feed and maintenance labor must be transported to the facility and the product returned. Siting decisions are complicated by complex, politically driven permit problems in many countries that prevents optimal locations for the farms.

In sites without adequate currents, heavy metals can accumulate on the benthos (seafloor) near the salmon farms, particularly copper and zinc.

Contaminants are commonly found in the flesh of farmed and wild salmon. Health Canada in 2002 published measurements of PCBs, dioxins and furans and PDBEs in several varieties of fish. The farmed salmonids population had nearly 3 times the level of PCBs seen in the wild population. The farmed salmonids population had more than 3 times the level of PDBEs seen in the wild population. The farmed salmonids population had nearly twice the level of dioxins and furans seen in the wild population. On the other hand, "Update of the monitoring of levels of dioxins and PCBs in food and feed", a 2012 study from the European Food Safety Authority, stated that farmed salmon and trout contained on average a many times lesser fraction of dioxins and PCBs than wild-caught salmon and trout."

A 2004 study, reported in *Science*, analysed farmed and wild salmon for organochlorine contaminants. They found the contaminants were higher in farmed salmon. Within the farmed salmon, European (particularly Scottish) salmon had the highest levels, and Chilean salmon the lowest. The FDA and Health Canada have established a tolerance/limit for PCBs in commercial fish of 2000 ppb. A follow up study confirmed this, and found levels of dioxins, chlorinated pesticides, PCBs and other contaminants up to ten times greater in farmed salmon than wild Pacific salmon. On a positive note, further research using the same fish samples used in the previous study, showed that farmed salmon contained levels of beneficial fatty acids that were two to three times higher than wild salmon. A follow-up benefit-risk analysis on salmon consumption balanced the cancer risks with the (n−3) fatty acid advantages of salmon consumption. For this reason, current methods for this type of analysis take into consideration the lipid content of the sample in question. PCBs specifically are lipophilic, so are found in higher concentrations in fattier fish in general, thus the higher level of PCB in the farmed fish is in relation to the higher content of beneficial n−3 and n−6 lipids they contain. They found that recommended levels of (n-3) fatty acid consumption can be achieved eating farmed salmon with acceptable carcinogenic risks, but recommended levels of (n-3) EPA+DHA intake cannot be achieved solely from farmed (or wild) salmon without unacceptable carcinogenic risks. The conclusions of this, from 2005 were that:

> "Consumers should not eat farmed fish from Scotland, Norway and eastern Canada more than three times a year; farmed fish from Maine, western Canada and Washington state no more than three to six times a year; and farmed fish from Chile no more than about six times a year. Wild chum salmon can be consumed safely as often as once a week, pink salmon, Sockeye and Coho about twice a month and Chinook just under once a month."

In 2005, Russia banned importing chilled fish from Norway, after samples of Norwegian farmed fish showed high levels of heavy metals. According to the Russian Minister of Agriculture Aleksey Gordeyev, levels of lead in the fish were 10 to 18 times higher than Russian safety standards and cadmium levels were almost four times higher.

Pollutants or Toxins Introduced by Pisciculturists

In 2006, eight Norwegian salmon producers were caught in unauthorized and unlabeled use of nitrite in smoked and cured salmon. Norway applies EU regulations on food additives, according to which nitrite is allowed as a food additive in certain types of meat, but not fish. Fresh salmon was not affected.

Kurt Oddekalv, leader of the Green Warriors of Norway, argues that the scale of fish farming in Norway is unsustainable. Huge volumes of uneaten feed and fish excrement pollute the seabed, while chemicals designed to fight sea lice find their way into the food chain. He says: "If people knew this, they wouldn't eat salmon", describing the farmed fish as "the most toxic food in the world". Don Staniford—the former scientist turned activist/investigator and head of a small Global Alliance Against Industrial Aquaculture—agrees, saying that a 10-fold increase in the use of some chemicals was seen in the 2016-2017 timeframe. The use of the toxic drug emamectin is rising fast. The levels of chemicals used to kill sea lice have breached environmental safety limits more than 100 times in the last 10 years.

Impact on Wild Salmonids

Farmed salmonids can, and often do, escape from sea cages. If the farmed salmonid is not native, it can compete with native wild species for food and habitat. If the farmed salmonid is native, it can interbreed with the wild native salmonids. Such interbreeding can reduce genetic diversity, disease resistance, and adaptability. In 2004, about 500,000 salmon and trout escaped from ocean net pens off Norway. Around Scotland, 600,000 salmon were released during storms. Commercial fishermen targeting wild salmon frequently catch escaped farm salmon. At one stage, in the Faroe Islands, 20 to 40 percent of all fish caught were escaped farm salmon. In 2017, about 263,000 farmed non-native Atlantic salmon escaped from a net in Washington waters in the 2017 Cypress Island Atlantic salmon pen break.

Sea lice, particularly *Lepeophtheirus salmonis* and various *Caligus* species, including *C. clemensi* and *C. rogercresseyi*, can cause deadly infestations of both farm-grown and wild salmon. Sea lice are naturally occurring and abundant ectoparasites which feed on mucus, blood, and skin, and migrate and latch onto the skin of salmon during planktonic nauplii and copepodid larval stages, which can persist for several days. Large numbers of highly populated, open-net salmon farms can create exceptionally large concentrations of sea lice; when exposed in river estuaries containing large numbers of open-net farms, many young wild salmon are infected, and do not survive as a result. Adult salmon may survive otherwise critical numbers of sea lice, but small, thin-skinned juvenile salmon migrating to sea are highly vulnerable. In 2007, mathematical studies of data available from the Pacific coast of Canada indicated the louse-induced mortality of pink salmon in some regions was over 80%. Later that year, in reaction to the 2007 mathematical study, Canadian federal fisheries scientists Kenneth Brooks and Simon Jones published a critique titled "Perspectives on Pink Salmon and Sea Lice: Scientific Evidence Fails to Support the Extinction Hypothesis". The time since these studies has shown a general increase in abundance of Pink Salmon in the Broughton Archipelago. Another comment in the scientific literature by Canadian Government Fisheries scientists Brian Riddell and Richard Beamish et al. came to the conclusion that there is no correlation between farmed salmon louse numbers and returns of pink salmon to the Broughton Archipelago. And in relation to the 2007 Krkosek extinction theory: "the data was used selectively and conclusions do not match with recent observations of returning salmon".

A 2008 meta-analysis of available data shows that salmonid farming reduces the survival of associated wild salmonid populations. This relationship has been shown to hold for Atlantic, steelhead, pink, chum, and coho salmon. The decrease in survival or abundance often exceeds 50%. However, these studies are all correlation analysis and correlation doesn't equal causation, especially when similar salmon declines were occurring in Oregon and California, which have no salmon aquaculture or marine net pens. Independent of the predictions of the failure of salmon runs in Canada indicated by these studies, the wild salmon run in 2010 was a record harvest.

A 2010 study that made the first use of sea lice count and fish production data from all salmon farms on the Broughton Archipelago found no correlation between the farm lice counts and wild salmon survival. The authors conclude that the 2002 stock collapse was not caused by the farm sea lice population: although the farm sea lice population during the out-migration of juvenile pink salmon was greater in 2000 than that of 2001, there was a record salmon returning to spawn in 2001 (from the juveniles in 2000) compared with a 97% collapse in 2002 (from the juveniles in 2001). The authors also note that initial studies had not investigated bacterial and viral causes for the event despite reports of bleeding at the base of the fins, a symptom often associated with infections, but not with sea lice exposure under laboratory conditions.

Wild salmon are anadromous. They spawn inland in fresh water and when young migrate to the ocean where they grow up. Most salmon return to the river where they were born, although some stray to other rivers. Concern exists about of the role of genetic diversity within salmon runs. The resilience of the population depends on some fish being able to survive environmental shocks, such as unusual temperature extremes. The effect of hatchery production on the genetic diversity of salmon is also unclear.

Genetic Modification

Salmon have been genetically modified in laboratories so they can grow faster. A company, Aqua Bounty Farms, has developed a modified Atlantic salmon which grows nearly twice as fast (yielding a fully grown fish at 16–18 months rather than 30), and is more disease resistant, and cold tolerant. It also requires 10% less food. This was achieved using a chinook salmon gene sequence affecting growth hormones, and a promoter sequence from the ocean pout affecting antifreeze production. Normally, salmon produce growth hormones only in the presence of light. The modified salmon does not switch growth hormone production off. The company first submitted the salmon for FDA approval in 1996. In 2015, FDA has approved the AquAdvantage Salmon for commercial production. A concern with transgenic salmon is what might happen if they escape into the wild. One study, in a laboratory setting, found that modified salmon mixed with their wild cohorts were aggressive in competing, but ultimately failed.

Impact on wild Predatory Species

Sea cages can attract a variety of wild predators which can sometimes become entangled in associated netting, leading to injury or death. In Tasmania, Australian salmon-farming sea cages have entangled white-bellied sea eagles. This has prompted one company, Huon Aquaculture, to sponsor a bird rehabilitation centre and try more robust netting.

Impact on Forage Fish

The use of forage fish for fish meal production has been almost a constant for the last 30 years and at the maximum sustainable yield, while the market for fish meal has shifted from chicken, pig, and pet food to aquaculture diets. This market shift at constant production appears an economic decision implying that the development of salmon aquaculture had no impact on forage fish harvest rates.

Fish do not actually produce omega-3 fatty acids, but instead accumulate them from either consuming microalgae that produce these fatty acids, as is the case with forage fish like herring and sardines, or consuming forage fish, as is the case with fatty predatory fish like salmon. To satisfy this requirement, more than 50% of the world fish oil production is fed to farmed salmon.

In addition, salmon require nutritional intakes of protein, which is often supplied in the form of fish meal as the lowest-cost alternative. Consequently, farmed salmon consume more fish than they generate as a final product, though considerably more preferred as food.

Hatch and Release

Another form of salmon production, which is safer but less controllable, is to raise salmon in hatcheries until they are old enough to become independent. They are then released into rivers, often in an attempt to increase the salmon population. This practice was very common in countries such as Sweden before the Norwegians developed salmon farming, but is seldom done by private companies, as anyone may catch the salmon when they return to spawn, limiting a company's chances of benefiting financially from their investment. Because of this, the method has mainly been used by various public authorities and nonprofit groups, such as the Cook Inlet Aquaculture Association, as a way of artificially increasing salmon populations in situations where they have declined due to overharvest, construction of dams, and habitat destruction or disruption. Unfortunately, negative consequences to this sort of population manipulation can occur, including genetic "dilution" of the wild stocks, and many jurisdictions are now beginning to discourage supplemental fish planting in favour of harvest controls and habitat improvement and protection. A variant method of fish stocking, called ocean ranching, is under development in Alaska. There, the young salmon are released into the ocean far from any wild salmon streams. When it is time for them to spawn, they return to where they were released where fishermen can then catch them.

Land-raised Salmon

Recirculating acquaculture systems make it possible to farm salmon entirely on land, which as of 2019 is an ongoing initiative in the industry. However, large farmed salmon companies such as Mowi and Cermaq were not investing in such systems. In the United States, a major investor in the effort was Atlantic Sapphire, which planned to bring salmon raised in Florida to market in 2021. Other companies investing in the effort include Nordic Acquafarms and Whole Oceans.

COMMERCIAL FISHING

Commercial fishing is the taking of fish and other seafood and resources from oceans, rivers, and lakes for the purpose of marketing them. In the early 21st century about 250 million people were directly employed by the commercial fishing industry, and an estimated one billion people depended on fish as their primary source of animal protein.

Fishing is one of the oldest employments of humankind. Ancient heaps of discarded mollusk shells, some from prehistoric times, have been found in coastal areas throughout the world, including those of China, Japan, Peru, Brazil, Portugal, and Denmark. These mounds, known as kitchen middens, indicate that marine mollusks were among the early foods of humans.

Archaeological evidence shows that humans next learned to catch fish in traps and nets. These ventures were limited at first to the lakes and rivers, but as boats and fishing devices were improved, humans ventured into sheltered coastal areas and river mouths and eventually farther out onto the continental shelves, the relatively shallow ocean plains between the land and the deeper ocean areas. In some shelf areas where seaweed was abundant, this was also incorporated into the diet.

Fishing technology continued to develop throughout history, employing improved and larger ships, more sophisticated fishing equipment, and various food preservation methods. Commercial fishing is now carried on in all types of waters, in all parts of the world, except where impeded by depth or dangerous currents or prohibited by law. Commercial fishing can be done in a simple manner with small vessels, little technical equipment, and little or no mechanization as in small local, traditional, or artisanal fisheries. It can also be done on a large scale with powerful deep-sea vessels and sophisticated mechanical equipment similar to that of other modern industrial enterprises.

Fishing for beluga sturgeon in the Volga River, Volgograd.

Peru, Fishing for anchovies off the coast.

Both algae and animals are taken from the sea. Two types of fish are caught: demersal, living at or near the bottom, although sometimes in mid-water; and pelagic, living in the open sea near the surface. Cod, haddock, hake, pollock, and all forms of flatfish are common demersal fish. Herring and related species and tuna and their relatives are examples of pelagic fish. Both demersal and pelagic fish can sometimes be found far from coastal regions. Other aquatic animals that may be the object of commercial fishery include, most notably, crustaceans (lobsters, spiny lobsters, crabs, prawns, shrimps, crayfish) and mollusks (oysters, scallops, mussels, snails, squid, octopuses). Certain mammals (whales, porpoises), reptiles (serpents, crocodiles), amphibians (frogs), many types of worms, coelenterates (coral, jellyfish), and sponges are also sought in commercial fishing. Most of these animals are legally regarded as fish in many countries.

Lobster fishing off the coast of Cape Breton, Nova Scotia.

Various algae are commercially obtained in both seawater and fresh water. Seaweed is harvested in the water or collected on the seashore. Algae play an important ecological role in many countries, not only as human food but also as fodder for cattle, as fertilizer, and as a raw material for certain industries.

Fisheries are classified in part by type of water: fresh water—lake, river, and pond—and salt water—inshore, mid-water, and deep sea. Another classification is based on the object—as in whaling, salmon fishing, and sponge fishing. Sometimes fisheries are classified according to the method of fishing employed: harpooning, seining, trawling, and lining.

While fisheries are considered renewable resources, overfishing has depleted fish and other seafood in many places and is a major threat to aquatic biodiversity. In addition, the use of less-selective fishing gear, such as gillnets or bottom trawls, results in substantial bycatch (the incidental catch of non-target species); some estimates state that bycatch may amount to as much as 40 percent of the global catch. The sustainable management of fisheries is key to both the health of aquatic ecosystems and the continued productivity of commercial fishing.

Food-gathering peoples first obtained fish and shellfish from the shallow water of lakes and along the seashore, from small ponds remaining in inundation areas, from tidal areas, and from small streams. Some authorities believe that in the earliest times fish were rarely caught because of the inadequacy of fishing gear. Shellfish, however, can be gathered easily by hand, and the prehistoric kitchen middens indicate their importance as a food source.

Bycatch, A bigeye thresher shark (Alopias superciliosus) hooked on a longline as bycatch.

In earliest times most foodstuffs were used at once and not stored, but as expanding populations increased food needs, techniques were developed for preserving fish by drying, smoking, salting, and fermentation. It became desirable to catch large quantities, and specialized equipment was devised. Individual fishing was replaced by collective efforts involving larger, more effective gear.

Fishing equipment and methods improved through the centuries, until bulk fisheries were established in Europe. Herring were caught in huge numbers in northern Europe in the Middle Ages. Cod fishery began on the Grand Banks of Newfoundland even before the Italian explorer John Cabot made his voyage there in 1497. Whaling with large fleets began in the 17th century, both in the Atlantic and in the South Pacific.

Before mechanization came to the fishing industry toward the end of the 19th century, sailing vessels developed to suit conditions and fisheries in different areas. The Grand Banks schooners were the peak of such developments. Sailing from New England, Nova Scotia, and Newfoundland, they fished cod during trips lasting up to six months, salting the catch for export to Europe, Africa, and the Caribbean. Individuals fished from small wooden dories, setting and hauling longlines by hand. Portuguese vessels also sailed annually to the Grand Banks, and a number still operated alongside modern steel vessels into the late 20th century. Smaller cutters and yawls worked around Europe using drift gill nets and setnets. Beam trawls were used extensively in the North Sea and English Channel, particularly for flatfish, being towed downwind under sail and then hauled back to the vessel's side.

As steam-driven winches came into use, fishing gear increased in size and weight. Steam gradually replaced sail for propulsion in the last quarter of the 19th century. In turn, the internal-combustion engine supplanted steam, although steam trawlers continued operating as late as the 1950s. Smaller craft became motorized in the early 20th century, and the inboard diesel engine became universally adopted—except for the smallest boats, on which gasoline-powered outboard engines remain common.

Larger catches could be obtained by increasing the number or the size of the fishing gear or both. Simple lines armed with one or a few hooks were replaced by longlines with thousands of hooks. Single small traps were combined into systems of hundreds, and pots were set in large quantities.

Nets were greatly enlarged; netmaking machines were invented that produced netting in large sheets. Mechanical netmaking brought replacement of the old local netting fibres (linen and hemp) with cotton and hard fibres. But all natural fibres, especially those of cellulose, begin to rot in time; thus, the introduction after World War II of rot-proof nets made of synthetic fibres represented a major advance. Mechanical netmaking remained unchanged for the most part, though for certain fishing gear the usual knotted netting was replaced by knotless netting.

In the beginning of the 1950s, mechanization took a great stride forward in purse seining when the power block was invented for hauling the gear. Another important hauling device was a power-driven drum to haul and store seine nets, gill nets, purse seines, and even the large trawl nets. The Japanese introduced drums in longline fishing for tuna. Another important innovation was the stern chute for stern trawlers, a development made possible by cooperation between naval architects and fishing-gear experts, which permitted large-scale mechanization of gear handling.

An era of rapid technical development in vessel design began with the British factory trawler experiment in the late 1940s, which demonstrated the great advantage of large stern trawlers that processed their catch on board. The idea was quickly developed by countries seeking to fish distant resources, and by the mid-1960s these large vessels (up to 100 metres [328 feet] long) were being operated by the Soviet Union, the United Kingdom, Japan, Poland, East Germany, and Spain.

Equivalent development occurred in the exploitation of the huge resources of small pelagic fish, mainly for conversion into fish meal. In the late 1940s small vessels, using hand-operated natural-fibre nets, fed small shore-based canning and fish meal plants. By the late 1960s large fleets of 25-metre (82-foot) purse seiners were supplying factory mother ships capable of handling up to 3,000 tons per day.

Concurrently, developing countries strove to introduce more modern fishing technology in order to boost protein supplies for their populations. Most rely heavily on artisanal fishing, using canoes or small boats with simple gear and often working off of open beaches. The introduction of outboard motors, larger boats, and synthetic nets enabled many countries to increase their catches significantly.

In the 40 years following World War II, the annual world fishing catch quadrupled. By the early 1970s, though, it had become apparent that such development was not limitless. Several of the largest resources of pelagic fish harvested by purse seiners suffered collapses generally blamed on overfishing. These included the northeast Atlantic herring, the South Atlantic pilchard, and the West African sardine and associated species. Severe declines in catches of stocks fished by fleets of factory trawlers caused such concern that coastal states pressed for protection of the resources off their shores. In 1972 Iceland became the first country to claim an extended fisheries limit of 80 km (50 miles) and, in 1975, 322 km (200 miles). Other countries followed suit, and in 1983 the Law of the Sea established an exclusive economic zone, or EEZ, of 322 km (200 miles), inside of which each country had exclusive right to the exploitation of marine life. An immediate result was the exclusion from many areas of high-performance long-distance foreign fleets, which were replaced by often less-efficient domestic coastal craft. For example, the British fleet of 168 distant-water trawlers disappeared within a few years, replaced by a fleet of compact coastal-type vessels.

Commercial fishing in areas outside EEZs are managed by Regional Fisheries Management Organizations (RFMOs) and their member countries. Members include bordering countries as well as more distant countries that fish in those waters; Japan, for example, has fleets in the Atlantic and is thus a member of the RFMOs that regulate the region. Many economically important fish species, such as salmon and tuna, are managed by specific RFMOs. Despite these efforts, illegal and unreported fishing is common in many areas, especially in the Indian Ocean.

The oil crisis of the 1970s increased fuel costs as much as 400 percent while fish prices rose by only about 80 percent. This forced many fuel-inefficient vessels, such as many of the U.S. Gulf shrimp trawlers, to tie up or transfer to other fisheries. Resulting development of fuel-efficient vessels, engines, fishing methods, and equipment—including applications of modern sail technology—depended thereafter upon the price of oil.

With the growing importance of managing fisheries to ensure maximum possible benefit from a particular stock, the work of fisheries scientists increased in importance. From a mainly descriptive science in the 19th century, the field evolved, especially after World War II, to develop sophisticated computer analyses based on mathematical models to predict the optimum yields available from fish populations. Still, fishing laws and regulations are difficult to enforce, and international waters are often subject to the tragedy of the commons. In the early 21st century it was estimated that a third of the world's fisheries were overexploited, and stocks of large fish had dwindled by an estimated 90 percent.

References

- Fishery, entry: newworldencyclopedia.org, Retrieved 31 March, 2019
- Artisanal-fisheries: weebly.com, Retrieved 14 July, 2019
- integrated sea lice management strategies – scottish salmon producers' organisation. Scottishsalmon.co.uk (2013-11-23). Retrieved on 2016-10-26
- Fisheries-fresh-water, en-ge: waterencyclopedia.com, Retrieved 17 May, 2019
- Stead, selina m.; laird, lindsay (14 january 2002). The handbook of salmon farming. Springer science & business media. Pp. 348–. Isbn 978-1-85233-119-1
- Marine-fisheries, energy-government-and-defense-magazines, environment: encyclopedia.com, Retrieved 19 April, 2019
- Commercial-fishing, technology: britannica.com, Retrieved 5 February, 2019

Fish Farming

CHAPTER 2

The raising of fish commercially in tanks or fish ponds usually for food is referred to as fish farming. It is also involved in the prevention of diseases in fish as well as their treatment. This chapter has been carefully written to provide an easy understanding of these facets of fish farming.

Fish farming refers to the commercial production of fish in an enclosure or, when located in a body of freshwater or marine water, in an area that is penned off from the surrounding water by cages or open nets.

A fish farm is similar to a fish hatchery in that both can contain 500,000 and more fish. But, a fish hatchery is designed to raise the fish only to a young age before they are released into the wild, usually to bolster the numbers of that species. In contrast, a fish farm is designed to raise the fish until they are a size and age that makes them the best commercial value. The fish are ultimately retrieved and sold, typically as whole or processed food.

Fish farming is the most common form of aquaculture, and commonly involves trout, salmon, tilapia, cod, carp, and catfish. For a species such as cod, whose numbers in the Grand Banks fishery off the east coast of the Canadian maritime provinces plummeted to near zero in the 1970s due to overfishing, and as of 2008 have yet to recover, the cod available from fish farming represents almost the sole source of the fish in North American markets.

The example of cod is cited as one of the advantages of fish farming. Raising fish under more controlled conditions that are possible in the wild avoids the problem of overfishing. As well, because an operation takes up relatively little space, feeding and care of the fish can be done under more controlled conditions, which is an economic advantage to those who own and run the facility.

However, fish farming is a controversial practice. For example, on the Canadian west coast, the farming of salmon typically uses species normally found in the Atlantic Ocean. The escape of fish to the wild does occur, and has created concern that the presence of the species in an environment that is unnatural to them could upset the marine ecology. Other concerns of fish farming are the overcrowding of fish, which can make them more susceptible to disease such as sea lice, and the use of antibiotics, which can also be released into the natural environment.

Fish farming is an ancient practice, dating back to about 2500 BC in China, when carp were raised in ponds and in artificial lakes created by receding floodwaters. Some of the motivations for fish farming in ancient China are shared with fish farm owners and operators in 2008. These include maximizing the food available from the resource; reducing the energy needed to search for, gather, and transport the food; making food production more predictable and less likely to be influenced by weather, predators, or other factors; and ensuring that the quality of the resource remains acceptable over time.

Evidence of fish farming also dates back at least 1,000 years in Hawaii, when rocks were added to existing reefs to create an artificial pond. The spaces between the rocks were large enough to let seawater circulate in and out, allowing nutrients to circulate in and waste out, but were too small to allow the fish to escape. The open net design of present-day fish farms follows this example.

Fish farming became more prevalent in Europe in the fifteenth century. The first known fish hatchery constructed in North America was built in the Canadian province of Newfoundland in 1889.

In the 1960s, fish farming expanded worldwide as some commercial fish stocks became less plentiful and a growing global population increased the demand for fish. As with factory farms—the landlocked facilities in which huge numbers of poultry and livestock are raised—economic incentives were provided to encourage the establishment of freshwater and marine water fish farms. In addition, corporations involved in the sales of fresh and processed fish products began to expand into fish farming as a way of ensuring supply, expanding their market, and trimming costs.

As practiced in China thousands of years ago, fish farming was efficient and sustainable. The numbers of fish were suitable for the space available and the population was managed so that the numbers of fish ready for harvesting did not decline over time. When done in such a sustainable way, fish farming can be a good strategy to supplement or even replace the fish caught from the wild.

However, the confinement of large numbers of fish in a small area can create problems. In an enclosed pond, one problem can be the accumulation of waste products. Aside from making the water less hospitable for the fish, the waste material can act as a food source for microorganisms known as algae. Combined with suitable water temperature and sunlight, the presence of the food source can lead to the rapid increase in the number of algae termed as algal blooms. The number of algae in blooms that occur in the open ocean can be so large that the growth is visible from orbiting satellites. In a confined pond, an algal bloom can use up much of the oxygen in the water, leading to the death of the fish.

When a fish farm is done in tanks (closed circulation type), it is essential to keep the water well oxygenated and to remove wastes. Bubbling air into the water as is done with a home aquarium can be one means of oxygenation. Alternatively, water can be cascaded from tank to tank in the fish farm complex, with oxygen being supplied as the water tumbles between tanks. Waste removal is typically done by continual removal of the water and

Impacts and Issues

Fish farming has become a very contentious practice, for a number of environmental reasons and for the adverse health effects it has on the farmed fish and possibly other species, including humans.

In a fish farm, the concentration of fish far exceeds that found in schools of fish in the wild—50,000 or more fish in an area of several acres in volume—with the possible exception of the spawning runs of west coast salmon. These crowded conditions reduce the free-swimming volume of each fish to about that of the average household bathtub. In such crowded conditions, the fish bump and rub against each other in the boundaries of the pens, which can produce cuts and scrapes. This increases the likelihood of infection and the development of diseases.

Species of sea lice that parasitize Coho and Atlantic salmon are especially troublesome. The sea lice attach to the fish and feed on tissue, which creates lesions and causes fluid loss from the affected fish. The confined fish become ill and can die. In addition, the sea lice can spread to wild salmon in the seas around fish farms when farmed salmon escape from the confinement, and also when the lice are washed away from the fish farm into the surrounding water. A 2001 survey of wild juvenile salmon migrating past fish farms in British Columbia found many more sea lice on the juveniles that had passed the farms than on those who had not yet passed by the facilities.

The escape of fish from fish farms is not a trivial and isolated event. Rips and breaks in the pen material and buffeting of the pens by storm-driven waves can lead to the escape of fish. In some cases, pens are designed with a net lid to reduce this possibility. Sometimes only a few fish escape. But mass escapes have occurred. For example, in January 2002, over 8,000 fish escaped from a fish farm in Clayoquot Sound, British Columbia. Worldwide in 2004, an estimated 2 million farmed fish escaped to the wild.

Once in the wild, the farmed fish have the potential to transfer disease to the wild population. An article in a December 2007 issue of Science documented declines in the population of wild Pacific salmon related to their decimation from sea lice transferred from farm populations of Atlantic salmon. The situation is so dire that the natural population could be reduced by 99% by 2015, which would be an economic disaster for the traditional salmon fishery and those employed by the fishery.

Antibiotics can be supplied to the food in an effort to control infections. As with land-bound factory farms, this practice encourages the development of antibiotic resistance among the surviving bacteria. These hardier varieties of the bacteria may pose a health hazard not only to the farmed fish, but to wild fish populations and to humans.

A fish farm releases a great amount of untreated sewage to the surrounding water. A study done in Clayoquot Sound calculated that the 700,000 fish housed in the facility that is the size of three football fields generate the daily equivalent amount of sewage produced by 150,000 people.

Elsewhere, the situation is not better. In China, for example, which produces approximately 70% of the world's farmed fish, fish farms can be concentrated together around large ponds. Analysis of the pond waters has revealed the presence of pesticides and other agricultural run-off, antibiotics, and cancer-causing compounds. The result is both an environment and food safety problem. In late 2007, the United States and China signed an agreement to permit more monitoring of the farms and the safety of the exported products.

Aquaculture makes use of local photosynthetic production (extensive) or fish that are fed with external food supply (intensive).

Extensive Aquaculture

Another issue is the risk of algal blooms. When temperatures, nutrient supply, and available sunlight are optimal for algal growth, algae multiply at an exponential rate, eventually exhausting nutrients and causing a subsequent die-off in fish. The decaying algal biomass depletes the oxygen in the pond water because it blocks out the sun and pollutes it with organic and inorganic solutes (such as ammonium ions), which can (and frequently do) lead to massive loss of fish.

Aqua-Boy, a Norwegian live fish carrier used to serve the Marine Harvest fish farms on the west coast of Scotland.

An alternate option is to use a wetland system, such as that used in the commercial fish farm Veta La Palma, Spain.

To tap all available food sources in the pond, the aquaculturist chooses fish species that occupy different places in the pond ecosystem, e.g., a filter algae feeder such as tilapia, a benthic feeder such as carp or [catfish, and a zooplankton feeder (various carps) or submerged weeds feeder such as grass carp.

Despite these limitations, significant fish farming industries use these methods. In the Czech Republic, thousands of natural and semi-natural ponds are harvested each year for trout and carp. The large Rožmberk Pond near Trebon, built in 1590, is still in use.

Intensive Aquaculture

Optimal water parameters for cold- and warm-water fish in intensive aquaculture	
Acidity	pH 6–9
Arsenic	<440 µg/l
Alkalinity	>20 mg/l (as $CaCO_3$)
Aluminum	<0.075 mg/l
Ammonia (non-ionized)	<O.O2mg/l
Cadmium	<0.0005 mg/l in soft water; <0.005 mg/L in hard water
Calcium	>5 mg/l
Carbon dioxide	<5–10 mg/l
Chloride	>4.0 mg/l
Chlorine	<0.003 mg/l
Copper	<0.0006 mg/l in soft water; <0.03 mg/l in hard water
Gas supersaturation	<100% total gas pressure (103% for salmonid eggs/fry) (102% for lake trout)
Hydrogen sulfide	<0.003 mg/l
Iron	<0.1 mg/l
Lead	<0.02 mg/l
Mercury	<0.0002 mg/l
Nitrate	<1.0 mg/l
Nitrite	<0.1 mg/l
Oxygen	6 mg/l for coldwater fish 4 mg/l for warmwater fish
Selenium	<0.01 mg/l
Total dissolved solids	<200 mg/l
Total suspended solids	<80 NTU over ambient levels
Zinc	<0.005 mg/l

In these kinds of systems fish production per unit of surface can be increased at will, as long as sufficient oxygen, fresh water and food are provided. Because of the requirement of sufficient fresh water, a massive water purification system must be integrated in the fish farm. One way to achieve this is to combine hydroponic horticulture and water treatment. The exception to this rule are cages which are placed in a river or sea, which supplements the fish crop with sufficient oxygenated water. Some environmentalists object to this practice.

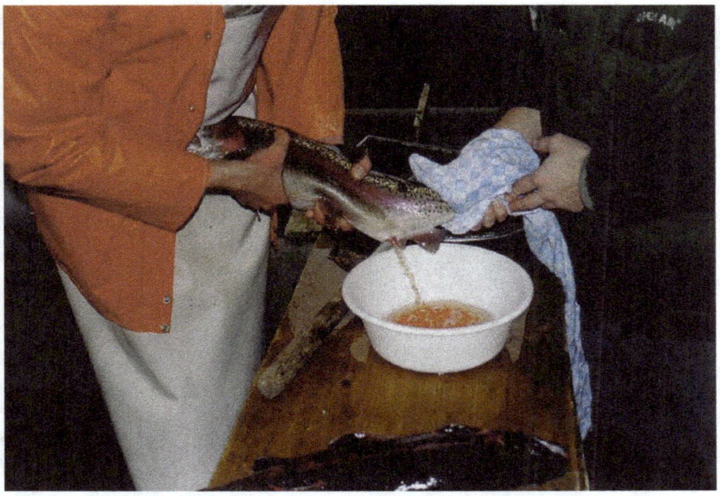

Expressing eggs from a female rainbow trout.

The cost of inputs per unit of fish weight is higher than in extensive farming, especially because of the high cost of fish feed. It must contain a much higher level of protein (up to 60%) than cattle feed and a balanced amino acid composition, as well. These higher protein-level requirements are a consequence of the higher feed efficiency of aquatic animals (higher feed conversion ratio [FCR], that is, kg of feed per kg of animal produced). Fish such as salmon have an FCR around 1.1 kg of feed per kg of salmon whereas chickens are in the 2.5 kg of feed per kg of chicken range. Fish do not use energy to keep warm, eliminating some carbohydrates and fats in the diet, required to provide this energy. This may be offset, though, by the lower land costs and the higher production which can be obtained due to the high level of input control.

Aeration of the water is essential, as fish need a sufficient oxygen level for growth. This is achieved by bubbling, cascade flow, or aqueous oxygen. *Clarias* spp. can breathe atmospheric air and can tolerate much higher levels of pollutants than trout or salmon, which makes aeration and water purification less necessary and makes *Clarias* species especially suited for intensive fish production. In some *Clarias* farms, about 10% of the water volume can consist of fish biomass.

The risk of infections by parasites such as fish lice, fungi (*Saprolegnia* spp.), intestinal worms (such as nematodes or trematodes), bacteria (e.g., *Yersinia* spp., *Pseudomonas* spp.), and protozoa (such as dinoflagellates) is similar to that in animal husbandry, especially at high population densities. However, animal husbandry is a larger and more technologically mature area of human agriculture and has developed better solutions to pathogen problems. Intensive aquaculture has to provide adequate water quality (oxygen, ammonia, nitrite, etc.) levels to minimize stress on the fish. This requirement makes control of the pathogen problem more difficult. Intensive aquaculture requires tight monitoring and a high level of expertise of the fish farmer.

Very-high-intensity recycle aquaculture systems (RAS), where all the production parameters are controlled, are being used for high-value species. By recycling water, little is used per unit of production. However, the process has high capital and operating costs. The higher cost structures mean that RAS is economical only for high-value products, such as broodstock for egg production, fingerlings for net pen aquaculture operations, sturgeon production, research animals, and some special niche markets such as live fish.

Controlling roes manually.

Raising ornamental coldwater fish (goldfish or koi), although theoretically much more profitable due to the higher income per weight of fish produced, has been successfully carried out only in the 21st century. The increased incidences of dangerous viral diseases of koi carp, together with the high value of the fish, has led to initiatives in closed-system koi breeding and growing in a number of countries. Today, a few commercially successful intensive koi-growing facilities are operating in the UK, Germany, and Israel.

Some producers have adapted their intensive systems in an effort to provide consumers with fish that do not carry dormant forms of viruses and diseases.

In 2016, juvenile Nile tilapia were given a food containing dried Schizochytrium in place of fish oil. When compared to a control group raised on regular food, they exhibited higher weight gain and better food-to-growth conversion, plus their flesh was higher in healthy omega-3 fatty acids.

Fish Farms

Within intensive and extensive aquaculture methods,numerous specific types of fish farms are used; each has benefits and applications unique to its design.

Cage System

Fish cages are placed in lakes, bayous, ponds, rivers, or oceans to contain and protect fish until they can be harvested. The method is also called "off-shore cultivation" when the cages are placed in the sea. They can be constructed of a wide variety of components. Fish are stocked in cages, artificially fed, and harvested when they reach market size. A few advantages of fish farming with cages are that many types of waters can be used (rivers, lakes, filled quarries, etc.), many types of fish can be raised, and fish farming can co-exist with sport fishing and other water uses.

Cage farming of fishes in open seas is also gaining popularity. Given concerns of disease, poaching, poor water quality, etc., generally pond systems are considered more simple to start and easier to manage. Also, past occurrences of cage-failures leading to escapes, have raised concern regarding the culture of non-native fish species in dam or open-water cages. In 2017, there was a massive failure of such cages at a commercial fishery in Washington state in Puget Sound, leading to the release of nearly 300,000 Atlantic salmon in non-native waters. This is believed to risk endangering the native Pacific salmon species.

Though the cage-industry has made numerous technological advances in cage construction in recent years, the risk of damage and escape due to storms is always a concern.

Semi-submersible marine technology is beginning to impact fish farming. In 2018, 1.5 million salmon are in the middle of a year-long trial at Ocean Farm 1 off the coast of Norway. The semi-submersible US$300 million project is the world's first deep-sea aquaculture project, and includes 61-meter (200 ft)-high by 91-meter (300 ft)-diameter pen made from a series of mesh-wire frames and nets, designed to disperse wastes better than more conventional farms in sheltered coastal waters, and therefore, be able to support higher fish packing density.

Copper-alloy Nets

Recently, copper alloys have become important netting materials in aquaculture. Copper alloys are antimicrobial, that is, they destroy bacteria, viruses, fungi, algae, and other microbes. In the marine environment, the antimicrobial/algaecidal properties of copper alloys prevent biofouling, which can briefly be described as the undesirable accumulation, adhesion, and growth of microorganisms, plants, algae, tube worms, barnacles, mollusks, and other organisms.

The resistance of organism growth on copper alloy nets also provides a cleaner and healthier environment for farmed fish to grow and thrive. Traditional netting involves regular and labor-intensive cleaning. In addition to its antifouling benefits, copper netting has strong structural and corrosion-resistant properties in marine environments.

Copper-zinc brass alloys are deployed in commercial-scale aquaculture operations in Asia, South America, and the USA (Hawaii). Extensive research, including demonstrations and trials, are being implemented on two other copper alloys: copper-nickel and copper-silicon. Each of these alloy types has an inherent ability to reduce biofouling, cage waste, disease, and the need for antibiotics, while simultaneously maintaining water circulation and oxygen requirements. Other types of copper alloys are also being considered for research and development in aquaculture operations.

In Southeast Asia, the traditional cage farming platform is called kelong.

Irrigation Ditch or Pond Systems

These use irrigation ditches or farm ponds to raise fish. The basic requirement is to have a ditch or pond that retains water, possibly with an above-ground irrigation system (many irrigation systems use buried pipes with headers.)

Using this method, water allotments can be stored in ponds or ditches, usually lined with bentonite clay. In small systems, the fish are often fed commercial fish food, and their waste products can help fertilize the fields. In larger ponds, the pond grows water plants and algae as fish food. Some of the most successful ponds grow introduced strains of plants, as well as introduced strains of fish.

Control of water quality is crucial. Fertilizing, clarifying, and pH control of the water can increase yields substantially, as long as eutrophication is prevented and oxygen levels stay high. Yields can be low if the fish grow ill from electrolyte stress.

Composite Fish Culture

The composite fish culture system is a technology developed in India by the Indian Council of Agricultural Research in the 1970s. In this system, of both local and imported fish, a combination of five or six fish species is used in a single fish pond. These species are selected so that they do not compete for food among them by having different types of food habitats. As a result, the food available in all the parts of the pond is used. Fish used in this system include catla and silver carp which are surface feeders, rohu, a column feeder, and mrigal and common carp, which are bottom feeders. Other fish also feed on the excreta of the common carp, and this helps contribute to the efficiency of the system which in optimal conditions produces 3000–6000 kg of fish per hectare per year.

One problem with such composite fish culture is that many of these fish breed only during monsoon. Even if fish are collected from the wild, they can be mixed with other species, as well. So, a major problem in fish farming is the lack of availability of good-quality stock. To overcome this problem, ways have now been worked out to breed these fish in ponds using hormonal stimulation. This has ensured the supply of pure fish stock in desired quantities.

FISH PRODUCTION IN INTEGRATED FARMING SYSTEMS

Integrated livestock-fish farming is a practice which links together two normally separate farming systems, whereby the livestock and fish become subsystems of a whole farming system. Emphasis focuses on an optimal waste or by-product utilization efficiency in which the waste of one subsystem (livestock) becomes an input to a second subsystem (fish). The integration of livestock with fisheriesaquaculture has received considerable attention lately with emphasis on the incorporation of animal manures as fertilizer and nutrient for promotion of natural feed in fish ponds. The recent surge of interest in integrated farming systems is due to the growing concern for maximizing productivity through optimum utilization of resources and to improve the diminishing per capita resources. The integration of duck and chicken with fish polyculture systems is amongst the most popular in Asian countries, followed by pig-fish and ruminants (cattle)-fish production systems. A more sophisticated integrated farming system is practiced in China; Chinese livestock-fish integrated farms may have complex interactions between livestock, crops and fish. In some Chinese farms, significant off-farm inputs to feed the fish such as aquatic macrophytes, snails and a wide variety of agricultural by-products are practised even though the pig may be the major livestock integrated with fish.

Promotion of livestock-fish integrated farming is viewed as a developmental strategy in overcoming a food crisis through enhancing waste and land-space utilization efficiency. The farming system improves space utilization and land-use conflicts in which the two subsystems essentially occupy all or part of the space required for an individual subsystem.

The Principles of Farming Technology of Fish Production In Integrated Farming Systems.

Guided by major findings on livestock-fish farming published in a preliminary set of principles defining the various aspects of the culture technology and promotion of fish production in integrated farming systems is suggested and outlined, as a basis for further deliberation. Principles pertaining to technology of livestock-fish farming are:

- Fish production in integrated systems is more complex than the conventional separate aquaculture system, requiring more knowledge and better management inputs.
- Integrated farming systems may vary in the degree of intensification of the livestock and fish subsystems, varying from extensive, semiintensive to intensive subsystems.
- Extensive subsystems utilize natural feed produced without international fertilization; semiintensive subsystems require fertilization to produce natural feed and/or supplementary feed but with a significant component of diet supplied by natural feed; and in intensive subsystems all the nutritional requirements are provided by artificial feed given to fishes with natural feed contributing little or no nutrition.
- Fish cultured in an integrated farming system with livestock benefit from a significant amount of the nutrition derived from natural food, which develops in the pond due to the fertilization by organic manures. This suggests the important role of pelagic algal-based food web which cultivates in fish biomass.
- The largest contribution of the manure to fish nutrition therefore, appears to be due to its fertilizing effect in the pond. Bacteria, breaking down the organic matter in the manure, release nutrients which lead to the production of phytoplankton and zooplankton.
- In a manure-fed fish pond, fish nutrition may also be derived from direct consumption of the manure. However, a detritus food web has a secondary role in fish biomass production biomass production, as compared with the algalbased food web.
- The direct nutritional value of manure for fish is by its content of spilled animal feed. A large proportion of the nutrients in livestock feed are not assimilated, but are voided in the excreta, particularly pig excreta.
- Over-fertilization with manure may lead to poor water quality of fish pond, particularly leading to depletion of dissolved oxygen and fish kills.
- Management of water quality is needed to overcome fish kills due to oxygen depletion and extreme fluctuation of dissolved oxygen levels. The strategy is to promote a growing biomass of phytoplankton which will generate sufficient oxygen to maintain relatively high dissolved oxygen have on a 24-hr basis. It is essential to maintain a positive net photosynthesis.

- The criteria for selection of fish species for stocking into manure-fed fish ponds should be based on the ability of fish species to:
    - Filter and feed on plankton (bacteria, phytoplankton and zooplankton).
    - Tolerate low levels of dissolved oxygen (< 2mg/1 minimum as defined by "Criteria for the Protection of Aquatic Life").
- An optimal stocking density of fish species is critical in attaining high cumulative fish yields and in reaching the upper carrying capacity of a manure-fed fish pond.
- Determination of and recommendation of the optimal stocking density of fish consider differences in local circumstances such as the fish species, manure-type and inputs to the pond, addition of other off-farm feed, and water quality of the pond.
- Ways to intensify fish production from integrated farming systems involve management inputs to:
    - Stock a higher initial fish biomass.
    - Harvesting the fish intermediately when the growth curve of stocked fish starts to slow down.
- There is a need for a more complex marketing system to handle the inputs and products from two subsystems as opposed to a single subsystem.
- Integrated farming on the one hand, enables the distribution of risk (both biological and economic), since two subsystems are involved as opposed to one in a single-commodity farming system; on the other hand, the failure of one subsystem can adversely affect the other.

Fish Species Suitability and Selection

The main objective of manuring is to produce a good growth of plankton, mainly phytoplankton, but also zooplankton and bacteria which are protein-rich natural for fish. On this basis, the most appropriate species of fish for a manure-fed pond are therefore those species which are able to filter and feed on plankton (bacteria, phytoplankton and zooplankton) from the water. The most common filter feeders are carps and mixed feeder on algae-detritus are certain species of tilapia. The most commonly raised carp species are the bighead carp Aristichthys nobilis Richardson and the silver carp Hypophthalmychithys molitrix Valenciennes. The carps are usually raised in polyculture with the grass carp Ctenopharyngodon idella and Indonesian carp Puntius gonionotus Linnaeus controls micro- and macro-vegetation at the water's edge, the Hoeven's slender carp Leptobarbus hoevenii Blk and the common carp Cyprinus carpio Linnaeus which is a bottom feeder. An important observation is that the detritus-feeding silver carp benefited most from the pond conditions, increasing its biomass share from an initial 5.8% to 20.7% from a series of investigation on the integrated farming systems in the Transkei, South Africa.

Polycultures of carps have been widely assumed to be the most efficient way to obtain high fish yields but in Taiwan, the percentage of tilapia in the pond has increased markedly at the expense of the Chinese carps. The most commonly raised tilapia species are the blue tilapia Oreochromis aureus Steindachner, the grey tilapia Oreochromis mossambicus Peters, the Nile tilapia Oreochromis niloticus Linnaeus or their hybrids. Tilapia are more suitable than carp for manure-fed ponds since the former are more tolerant to low levels of dissolved oxygen than the latter.

Air-breathing silver-striped catfish Clarias spp., are also integrated with pig farms. The detritus-feeding grey mullet Mugil cephalus L. may be excellent substitutes for the silver carp in integrated farming systems under saline conditions.

Fish Yields and Production Rates in Integrated Farming Systems

There are few data on fish yields from ponds which received livestock manure (and spilled livestock feed). The fish yields in integrated livestock-fish farming systems reported in the literature vary greatly, with the actual and extrapolated equivalents ranging from 1.5 ton/ha/year to 18 ton/ha/yr. The variation in fish yield is due:

- The animal sub-system and its major manure input to the pond,
- The fish species and stocking rates,
- Intensification of culture management inputs e.g. addition of other off-farm feed.

Different Types and Amounts of Manure Inputs

The rationale behind integrating fish with livestock is the large amount of nutrients (N-P-K) present in the feed that is recovered in the manure, with possible proportions of 72–79% nitrogen, 61–87% phosphorus, and 82–92% potassium. These act as fertilisers in fish ponds to produce plankton which comprise high-protein natural food for certain species of fish.

Nitrogen and phosphorus are the nutrients most likely to be limiting for plankton growth in the pond but fish yield is probably more directly correlated to manure nitrogen content since nitrogen is more volatile than phosphorus. Based on the nitrogen content of the different manures, it was possible to estimate the stocking density of different livestock (laying duck, dairy cow, pigs and buffalo) in order to produce the same fish yield of 174.7 kg/200 m^2/year in 13 family-level farms. Considering the livestock's manure characteristics (total live weight, age of livestock, feed characteristics of livestock, climate and management of livestock subsystem) and the varying N contents (% of the manure production/animal/year, i.e. 2.6 % for laying duck manure, 1.9 % for pig, 2.2 % for dairy cow and 1.1 % for buffalo, the estimated number of laying duck, pig, dairy cow and buffalo required to the same fish yields are 1335, 410, 40, and 85 individuals respectively. Prinsloo and Schoonbee noted the comparative effectiveness of the manure on the development of organisms in the food web and promotion of biological activity in fish ponds are : duck manure > pig manure > raw chicken manure > cattle manure > sheep manure.

Manure may contain up to 25% crude protein, but more than half of this is usually non-protein nitrogen (e.g. uric acid) which is not assimilated by fish; therefore manure itself is a poor feed for fish. Manures also contain less energy than conventional pelleted feeds. Though many species of fish consume manure directly, it is a low quality feedstuff compared to conventional pelleted feed and natural plankton.

Another advantage of manure utilization in fish ponds is that since carbon forms about 50% of the biomass of plankton, the high organic content of the manure is an important source of carbon which is released by bacterial respiration. However, depending on the productivity and quality of water and soil, the effectiveness of livestock manure (enriched with cellulose-rich organic matter) in acting as nutrient sources for fish growth and in raising fish yield varies. Two possibilities have been observed in Dor, Israel:

- Manure would improve fish yields only under conditions of water and soil which are poor in essential nutrient minerals (e.g. carbonates or organic matter), or of a low pH, fish ponds receiving very "soft" waters deficient in minerals, or ponds with a low primary production.
- Manure may be ineffective in raising fish yields above the rates achieved by daily applications of mineral (N-P-K) fertilisers, under conditions when the fish pond is receiving daily applications of mineral fertilisers, or where fish ponds are highly productive with very fertile soil naturally-enriched with nutrient minerals and carbonates, or fish ponds with nutrient-rich waters resulting in a high concentration of algalbased detrital organic matter.

Fish Species and Animal Stocking Rates

Conservatively, 5 tons of tilapia/ha/year is attainable for integration with 150 pigs. Higher yields of about 15 tons/ha/yr of the air-breathing silver-striped catfish can be expected from integration with about 300 pigs. However, lower yields of carps (less than 5 t/ha year) should be expected and they should be integrated with fewer pigs since carps are more sensitive to low levels of dissolved oxygen than tilapia.

Intensification of Management Inputs

Yields of fish in integrated farming systems can be considerably increased through intensification involving an increase in stocking density of fish, supplementary energy-rich fish feed (cereal or rice bran and complete pelleted feed) and aeration of pond water.

A 5 tons of tilapia/ha/year is a conservative fish production figure for integration with 150 pigs in Thailand. Higher tilapia yields of 15 tons/ha/year have been reported for integrated systems with pigs but with the fish also fed soyabean cake and rice bran and pelleted feed for the last 2 months of growth, and the pond aerated with several paddle-wheels.

Ways to Intensify Fish Production in Integrated Farming System

It is possible to intensify fish production further by the addition of an energy-rich feed such as cereal or rice bran, and still further by adding complete pelleted feed. To ensure increased fish yields with increased levels of feed, it may be necessary to increase the number of fish stocked in the pond. Dissolved oxygen levels may fall to low levels with intensification, so it may be necessary to aerate the water with a mechanical aerator. Yields of 1.5 – 4.7 tons/ha/year were reported from commercial duck/fish integrated farms in Hong Kong; besides duck manure and spilled duck feed, the fish were fed corn meal, rice or wheat bran, oil cakes and bean fragments. Much higher yields were reported from commercial farms in Taiwan where the fish were fed rice bran and soybean cake: A moderately-low production of 7.4–8.7 tons/ha/year were increased to 13.5–18.0 t/ha/year when the fish stocking density was increased, when the fish were fed pelleted feed during the last two months of the culture period, and when water was aerated by several sets of paddle wheels.

Another method of intensification of fish production is by stocking a higher initial fish biomass and by intermediate harvesting when the fish growth curve starts to slow down. Such a management strategy should lead to a considerably higher cumulative fish yield. Some integrated duck/fish farms in Hong Kong produce two or more crops of fish each year by initially stocking fish of different sizes, followed by repeated harvesting and stocking. Such method to intensify fish production is by manipulation of the fish stock to avoid both the initial slow rate of increase of the fish biomass (due to a low initial fish biomass) and the final slow rate of increase of fish biomass (due to the fish biomass approaching the carrying capacity or maximum biomass of fish that the pond can support under a given set of environmental conditions).

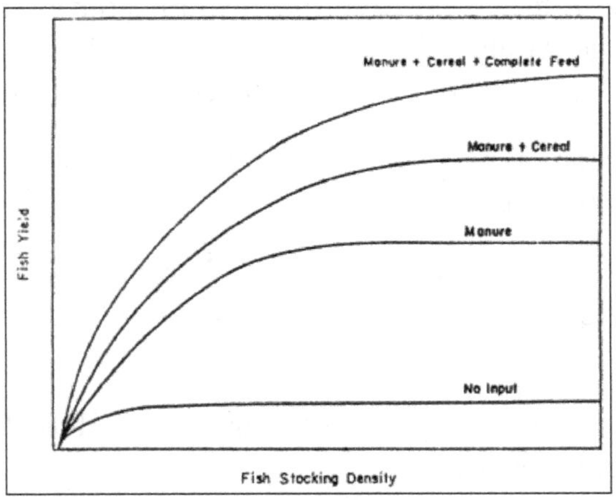

The relationship between fish stocking density and fish yield as a function of various types of pond nutritional inputs.

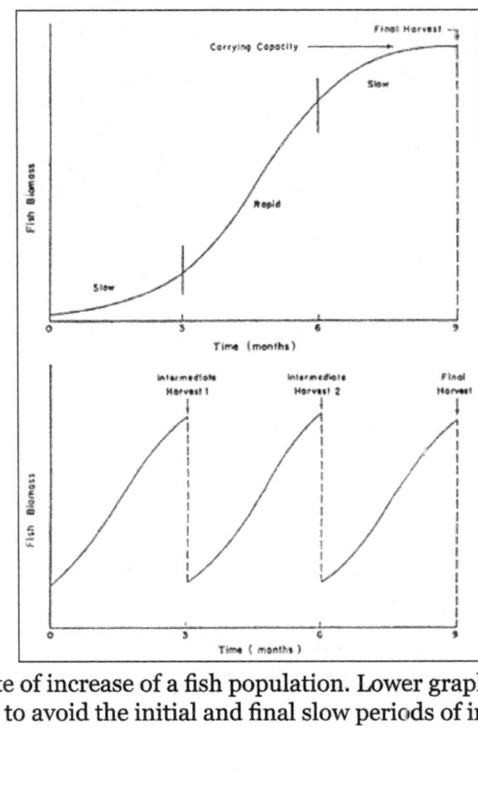

Upper graph normal rate of increase of a fish population. Lower graph, a higher initial stocking and intermediate harvesting to avoid the initial and final slow periods of increase in weight, respectively.

Advantages of Fish Farming Business

There are many advantages of starting commercial fish farming business and below given are some of the main benefits:

- As per the demand, commercial fish farming business allows for large supplies of fish. Catching fish from the wild cannot always fulfill the demand of consumers hence in such cases commercial fish farming can meet up this demand.
- You can easily raise the fishes in tanks until they are ready for selling or marketing and they don't need the wide capture of wild fish. Therefore commercial fish farming helps in preserving natural ecosystems.
- If compared to the wild fishes, some of the farm raised fish species are more healthy and delicious. Fishes are generally fed a wide variety of nutrient and protein enriched foods or pellets on commercial fish farms. So farm fish become healthier than the wild fish.
- There are indeed various types of fish species all over the world. So you can select your desired species for your fish farming business.
- We also know that fish is very popular as food all around the world thus there is already an established fish market available for you. And you don't really have to worry about where to sell them.
- You can start your fish farming business in large or small scale production; it all depends upon the resources available with you.

- And if you don't have sufficient funds or money to start this business, then you can also apply for bank loans.
- You can also fulfill your daily family nutritional demands through small-scale fish farming in tanks or ponds.

Last but not the least fish farming business is a great source of employment. It is important to mention that over 1 billion people all over the world depend on fishes as their primary protein source. And most of them are directly or indirectly involved with fish farming or fish products business. Consequently, fish farming generates a good income and employment source for these people. Moreover, global fish exportation business at present is earning more money every year, than any other food commodity.

DISEASE PREVENTION AND TREATMENT IN FISH FARMING

Fish diseases may cause severe losses on fish farms through:

- Reduced fish growth and production;
- Increased feeding cost caused by lack of appetite and waste of uneaten feed;
- Increased vulnerability to predation;
- Increased susceptibility to low water quality;
- Death of fish.

While it may be difficult to avoid fish diseases completely, it is better to try to prevent their occurrence rather than to allow them to develop and then attempting to cure them once they start to cause problems. To cure a fish disease is much more difficult and generally calls for the services of a specialist. By the time proper treatment can be organized, the disease may have become more serious. In some cases surviving fish are so weakened that effective treatment becomes difficult.

However several simple and effective treatments can be used, either for prevention or early control of disease before it becomes too serious. In all cases, you will have to handle chemicals and know about their dangers and their use.

Main Causes of Disease in Farm Fish

There are several causes of disease that may affect the fish directly or may continue to cause disease problems. Basically, any factor which causes stress or difficulty to the fish decreases its resistance to disease and increases the chance of disease problems occurring.

The three main causes of disease are:

- Improper feeding.
- Stress through extreme or toxic condition.
- Attack by disease organisms.
    - Fish are not fed properly: nutritional diseases become more frequent as the culture system becomes more intensive and the fish obtain less of their nutrients from natural food organisms.
    - Fish are stressed by being exposed to an extreme or a toxic condition.
- Rough and/or excessive handling, for example when harvesting or sorting/grading.
- Overcrowding and/or behavioral stresses, for example in storage or transport.
- Unsuitable water temperature.

- Lack of dissolved oxygen.
- Changes in pH towards extreme values.
- Presence of toxic gases such as ammonia or hydrogen sulphide.
- Toxic factors in artificial food such as particular chemicals in certain plant foodstuffs (saponin, gossypol, etc.), fungal toxins in stored foods, and pesticide residues.
- Pollution of the water by agricultural or industrial chemicals, sewage effluents, heavy silt loads.
    - Fish are attacked by successful disease organisms, either externally on the skin, gills or fins, or internally in the blood, digestive tract, nervous system, etc.

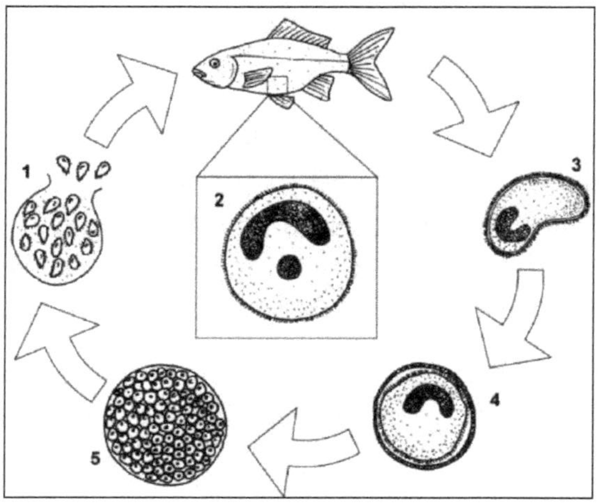

Life cycle of lchthyophthirius multifilis responsible for the white-spot disease.

This disease may spread rapidly from one fish to another through water and pond bottom infections which makes disease control very difficult.

Disease Caused by Living Organisms Develop in Fish Ponds

The development of disease caused by living organisms is encouraged by any condition which places the fish under stress.

Disease risks become even greater when fish undergo combined stresses, for example handling when the water temperature is below normal or overcrowding in low dissolved oxygen conditions.

Other factors on the fish farm may also be responsible for the survival and propagation of disease organisms, making disease control much more difficult such as:

- The presence of diseased wild fish.
- The presence of intermediate hosts such as snails and fish-eating birds, necessary for completing the life cycle of the disease organism.
- The introduction of disease organisms through contaminated inputs such as food, trash fish or processing wastes, for example imported eggs, juveniles, or broodstock, and water from an upstream pond or farm.

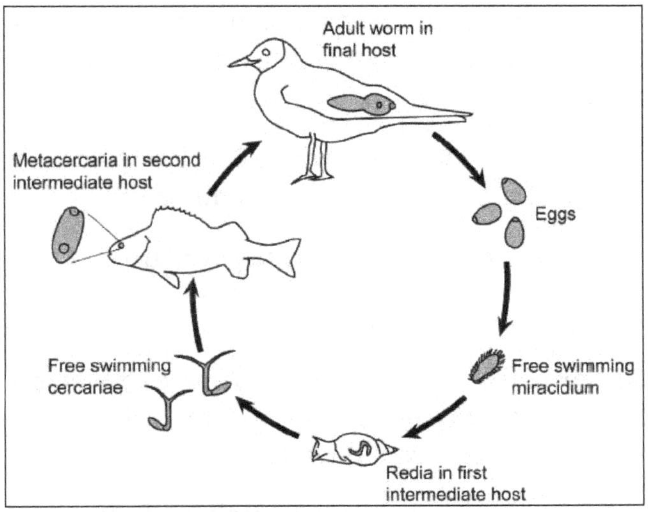

Life cycle of Diplostomum spathaceum.

Preventing Diseases through Good Management

Disease prevention on the fish farm is better than cure. All efforts should thus be directed to applying good management practices.

Be particularly vigilant about the following points:

- Ensure good water quality: sufficient supply, with adequate dissolved oxygen concentration and free of pollution.
- Keep the pond environment healthy: control silt,control plants,keep a healthy balance of phytoplankton and zooplankton, and exchange water if needed. If necessary, use mechanical aeration. Disinfect the pond regularly.
- Keep the fish in good condition:control stocking density. Keep different sizes or sexes separate if necessary to control fighting. Ensure good food supply. Handle the fish properly, especially during harvesting and sorting/grading. Care for your fish during storage and transport.
- Prevent the entry of disease organisms from outside your farm:
    - Control wild fish by using filters and screens and regularly eradicate them from canals and ponds.
    - Disinfect all fish stocks imported from outside as eggs, juveniles or adults.
    - Be careful when using trash fish or processing wastes as supplementary feed; if possible boil the raw material for at least 30 minutes or use it for compost or silage feeds; if natural food supplies are limited, add vitamin* supplements to the cooked food to ensure its quality.
    - Increase vigilance: If you have to use water downstream from a neighbouring fish farm, use screens to control escaped fish.

  - For a hatchery it is safest to use spring or well water, free of disease organisms; it is also useful for rearing small fry; alternatively, consider a sand filter to help remove smaller disease organisms.
  - Enclose hatchery and nursery areas with a fence to control access; use footbaths and protective clothing if necessary to limit contamination.
- Prevent the spread of disease organisms within your farm:
  - Control fish-eating predators, particularly birds and mammals.
  - Disinfect ponds regularly to kill both the disease organisms and their intermediate hosts; keep different age groups of fish separate; disinfect breeding ponds well and, if possible, remove broodstock from them as soon as spawning has taken place.
  - Use diversion ponds with parallel flow if possible; if your ponds are arranged in series, it is best to have the water flow from the ponds with the less infected and more sensitive, youngest fish into the ponds with the oldest fish (more infected and less sensitive).
  - Disinfect juveniles before stocking them in clean fattening ponds; treat broodstock before using them for propagation in breeding ponds.
  - If a disease breaks out on your farm, remove dead or dying fish from the ponds as quickly as possible, at least daily, and do not disturb and stress remaining fish excessively.
  - Bury diseased fish with quicklime away from the ponds; carefully treat infected ponds and disinfect all equipment that has come in contact with them.
  - In a hatchery have separate equipment for handling small and large fish, if possible keeping one set of hand nets, buckets, etc. for each tank or pond.
  - Use disinfectant bins for routine disinfection of equipment, and clearly mark the equipment accordingly.

Common Chemicals and their use on the Fish Farm

Most chemicals used for controlling disease organisms are toxic and/or irritant for the skin and respiratory tract. Many chemicals can cause serious health problems if swallowed or absorbed through the skin. You should therefore handle chemicals with care, mark them clearly and store them safely, away from children in particular.

When handling chemicals take at least the following precautions:

- Protect your hands by wearing rubber gloves; if possible wear overalls and boots for general protection.
- Work in a well-ventilated area.
- Handle chemicals carefully, avoiding splashing or spillage and, if accidentally splashed, wash off immediately.

- Avoid inhaling dangerous vapours.
- Clearly mark all containers, equipment and protective clothing used for storing and handling chemicals; unless they can be thoroughly and safely washed out, do not use them for other purposes; store them safely when not in use.
- Thoroughly wash your hands after use, and particularly before touching any food.

Buying and Storing Chemicals

Most chemicals degrade with time. When buying chemicals ensure that they are still of good quality. Check their expiration or "sell by" date, if any. Check also whether they have been stored properly. Clearly indicate on each container the date of purchase.

For storing chemicals, use a cool, dark, dry and lockable room. Some chemicals may require refrigerator storage; check labels or handling instructions. Generally, chemicals deteriorate more quickly in warm conditions. Keep good records and ensure a good rotation of the stocks.

Determining the Strength of Chemicals and their Solutions

Treatment chemicals are costly and can, in the wrong dosage, be toxic to fish. It is therefore essential to know and understand how the strength of these chemicals is expressed and how dosages can be calculated. You should become familiar with concentrations of chemical solutions, treatment units and their conversion values. You will then be able to avoid wasting chemicals and losing fish.

Treatment chemicals contain one or more toxic ingredients to disable or kill disease organisms. These are called the active ingredients (AI). The amount of active ingredient contained depends on the chemical. The AI concentration is expressed in percent of the total weight or volume of the chemical.

Example:

- Common salt contains 100 percent active ingredient.
- Chlorine bleach powder contains 33 percent Al (chlorine).
- Wescodyne contains 1.6 percent AI (iodine).
- Roccal contains from 10 to 50 percent Al (benzalkonium chlorides).

The usual practice is to dilute solid or liquid chemicals in water and prepare:

- A stock solution, which can be kept for some time, and be diluted as required to a working solution or a treatment dosage (also used when you need to dose very small amounts of chemical).
- A working solution, which is used directly once prepared.
- A treatment dosage, which is the concentration to which fish are exposed for treatment.

These solutions are prepared to a recommended strength or concentration, expressed either as:

- The concentration of the chemical required (e.g. diluted in water, or mixed with feed).
- The concentration of active ingredient required.

Be very careful to make sure whether it is the chemical or the active ingredient which is being referred to. Examples are given later for making up these solutions. The concentration of material (chemical or active ingredient) in a solution may be expressed in various ways:

- As the weight present in the solution volume, such as milligrams per litre solution (mg/l), grams per cubic metre (1000 l) solution (g/m^3), grams per 25 l solution (g/25 l).
- As a volume present in the solution volume such as millilitres per litre solution (ml/l), millilitres per cubic metre solution (mI/m^3), millilitres per 10 l solution (ml/10 l).

In practice, you may consider that:

- 1 ml water weighs 1 g
- 1 l water weighs 1 kg
    - As a percentage, the number of parts (normally by weight) of material in 100 parts of solution.

Example:

A 1.5 percent solution of salt contains 1.5 g salt per 100 g = 100 ml solution or 15 g/I; a 5 percent solution of chlorine bleach powder contains 5 g chlorine per 100 g = 100 ml solution or 50 g/I.

- In parts per million (ppm), the parts of material present in one million parts of solution; by weight this is equivalent to mg/l, g/m^3, or by volume to ml/m^3 of solution.

Example:

A 1000 ppm solution of chlorine bleach powder is in fact a 1000 mg/l solution; this amounts to 1 g chlorine per litre; a 160 ppm solution of formalin contains 160 mg/m^3 of formalin; this is equivalent to 0.16 ml formalin per litre solution.

- As a ratio, the parts of solution per each part of Al, such as a 1: 4 000 solution where for example each ml Al is present in 4 000 ml solution (1 ml/4 l).

Table: To obtain a ppm value from a ratio, multiply the ratio by 1 000 000 as shown in the chart below.

Ratio	ppm	Ratio	ppm
1: 500	2 000	1: 10 000	100
1: 1 000	1 000	1: 15 000	67
1: 4 000	250	1: 100 000	10
1: 5 000	200	1: 200 000	5
1: 6 000	167	1: 500 000	2

Calculating the Amount of Chemical to be used

You may wish to make a stock solution, a working solution or a treatment dosage of the chemical concerned. The concentration required will be determined by the type of chemical and treatment used, and the amounts used will depend on the concentration and the overall volume of the container, tank or pond involved.

Calculations of the amount to be used depend on the way the dosage is expressed and on the percentage of Al present in the chemicals.

If you are using a chemical with 100 percent Al, proceed in one of the following ways.

- When the concentration is given as a weight or volume of Al per volume of solution, multiply this concentration by the total volume of water to be treated.

Example:

- A concentration of 3 mg/l malachite green is required as a treatment dosage. Your tank contains 2 m^3 = 2000 l. You will need 3 mg/l x 2000 l = 6000 mg = 6 g malachite green.
- A stock solution of 1 g/I copper sulphate is to be made, to use in a 5-l drum. You will need 1 g/I x 5 l = 5 g of copper sulphate.
    - When the concentration is given as a percentage, multiply this concentration (expressed as a decimal number) by 1000 times the water volume (in l) to obtain the amount of chemical in ml or g.

Example:

Recommended treatment dosage is 2 percent common salt. Your plastic barrel contains 30 l water; you will need 0.02 x 1000 x 30 l = 600 g salt.

- When the concentration is given in parts per million (ppm), multiply this concentration by the water volume (in l). Divide the result by 1000 to obtain the amount of chemical in ml or g.

Example:

Recommended dosage is 100 ppm copper sulphate. Your trough contains 500 l water. You will need (100 ppm x 500 l) ÷ 1000 = 50 g copper sulphate.

- When the concentration is given as a ratio, multiply this concentration by 1000 times the water volume (in l) to obtain the amount of chemical in ml or g.

Example:

Recommended concentration is 1: 6000 formalin; your tank contains 2000 l water; you will need (1 ÷ 6000) x 1000 x 2000 l = about 333 ml of formalin.

If you are using a chemical that contains less than 100 percent active ingredient, divide the result obtained from the above calculations by the percentage of Al (expressed as a decimal) in the chemical.

Example:

Recommended dosage of an insecticide containing 80 percent Al is 0.5 ppm Al. Your pond contains 50 m^3 = 50000 l water:

- Calculate amount of Al required as (0.5 ppm x 50000 l) ÷ 1000 = 25 g;
- Calculate amount of chemical required to provide this as 25 g ÷ 0.80 = 31.25 g.

In the next sections, you will learn how to apply the above calculations on your fish farm. To do this, you need to know:

- How to measure solid and liquid chemicals accurately?
- How to measure water volumes?
- How to use stock solutions and working solutions?

Measuring Chemicals

To weigh chemical powder, crystals, etc., you require a balance with good accuracy within a particular range.

For small quantities, you may use, for example, a simple microbeam balance or a small spring balance, with good accuracy within the range of 1 to 125 g.

For larger quantities, you will need a good beam or spring balance similar to the one used to weigh fish, accurate within the range of 100 g to a few kilograms, depending on the volumes of water you plan to treat.

For quantities smaller than 1 g, use stock or working solutions.

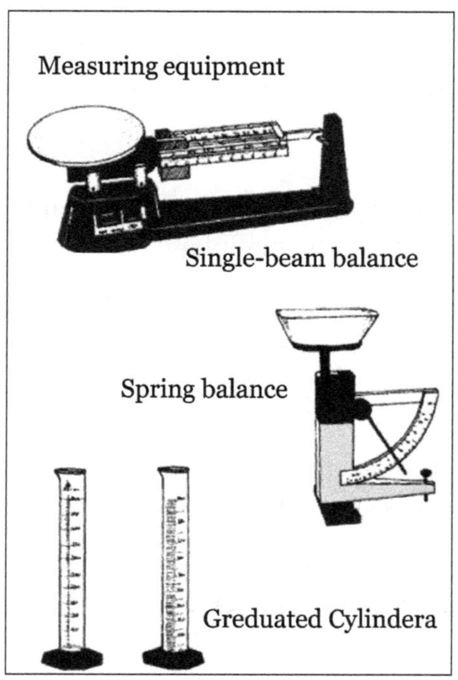

To measure volumes of liquid chemicals, you require a series of graduated plastic cylinders, with a typical range of capacity from 10 ml to 500 ml. The smaller the individual capacity of the cylinder, the more accurate the measurement.

You can get this equipment, either from school, medical or laboratory suppliers or from photographic supply stores.

For small volume measurements of one or a few millilitres, it is best to use either a medical syringe (capacity 2 to 5 ml) or a graduated or bulb-type pipette. Obtain these from a local pharmacy, drugstore or medical office.

For volumes smaller than 1 ml, use stock or working solutions.

Determining Water Volumes

The volume of water in a pond by multiplying its surface area by the average water depth. Keep in mind that both these values may change as the pond ages. Be sure to adjust your estimate of water volume accordingly before treating any pond.

To determine the volume of other structures or of various containers, proceed as follows:

- For circular tanks or drums, multiply area of base (in m^2) by water depth (in m). Calculate area of base as 3.14 x (D2 ÷ 4) where D (in m) is the inside diameter of the base.
- For a rectangular trough/tank or container, multiply base area (in m^2) by water depth (in m). Calculate base area as inside length (in m) times inside width (in m) of base.
- For rectangular tanks with sloping sides, use average width and length (at half water depth) to calculate area.
- For fully filled barrel-shaped containers, calculate the inside midsection area and the base area. Get the average value of the two (in m^2) . Multiply this by the water depth (in m).
- For circular tanks or drums with sloping sides, use the average area as for barrel-shaped containers, using the diameter at the base and at the upper water level. Multiply the average area by the water depth.
- For partially filled barrels:
    - When filled below the mid-section, treat as a sloping-sided drum or tank as above;
    - When filled above the mid-section, calculate volume up to mid-section as above and add the volume from mid-section to water level, also calculated as above.
- For small containers or those with an irregular shape: determine water volume by filling them up to the desired water depth with a container of known volume such as a bottle, a tin or a bucket.

It may he helpful to mark, using paint or indelible pen, the water levels corresponding to the water volumes you normally use. Good records will allow you to dose quickly and accurately for specific water volumes.

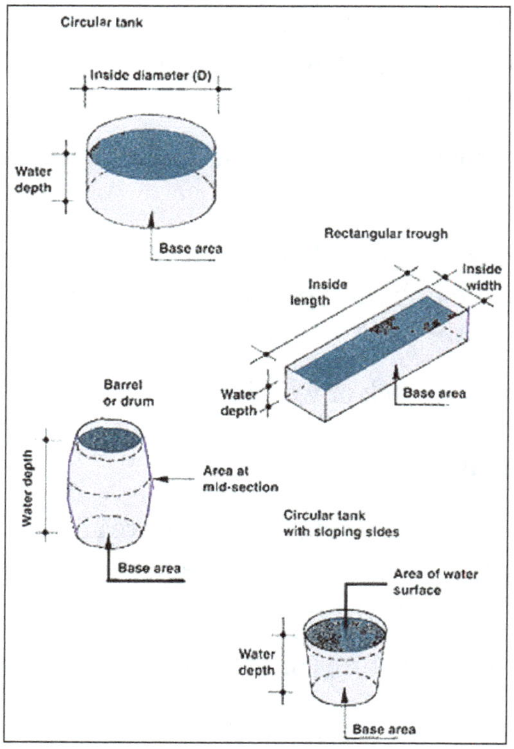

Example:

- A circular tank has an interior diameter D = 3 m and a water depth of 0.80 m.
- The volume of water it contains is 3.14 x [(3 x 3 m) ÷ 4] x 0.80 m = 5.652 m^3.
- A hatchery trough has interior dimensions of 1.5 x 0.4 m.
- Water depth is 0.25 m. Water volume is 1.5 x 0.4 x 0.25 m = 0.15 m^3 = 150 l.
- A metal drum has an interior diameter of 0.58 m. Water depth is 0.70 m.
- Water volume is 3.14 x [(0.58 x 0.58 m) ÷ 4] x 0.70 m = 0.185 m^3 = 185 l.
- A plastic barrel has a mid-height diameter of 0.50 m and a base diameter of 0.38 m (inside dimensions). Water depth is 0.75 m. Mid-section area is 3.14 x [(0.50 x 0.50 m) ÷ 4] = 0.196250 m^2. Base area is 3.14 x [(0.38 x 0.38 m) ÷ 4] = 0.113354 m^2. The average cross-section area is (0.196250 m^2 + 0. 113354 m^2) ÷ 2 = 0.154802 m^2 = 0. 155 m^2. The water volume is 0.155 m^2 x 0.75 m = 0.11625 m^3 = 116.25 l.

Using Stock and Working Solutions

The commonly available balances and measuring cylinders are not accurate enough to measure very small amounts of chemicals such as fractions of a gram or millilitre. It is best then to prepare a stronger stock solution, from which the final solution is prepared. The stock solution can usually be stored, to use as required. In some cases, you can dilute the stock solution to a working solution, which can be used for a short time, for example during one or two days, to be diluted as needed.

Stock solutions and working solutions are usually made up to concentrations which make them convenient to use, for example:

- For a stock solution of 1 g (1000 mg) of chemicals per litre, 1 ml contains 1 mg. Thus 1 ml mixed in 1 litre gives a solution of 1 mg/l, whereas 20 ml mixed in 1 litre gives a solution of 20 mg/l, etc.;
- For a working solution of 10 mg/l (for example made by mixing 10 ml of the previous stock solution in 1 litre of water), 100 ml contains 1 mg and 50 ml contains 0.5 mg of chemical. Thus 50 ml of working solution mixed in 1 litre of water gives a chemical concentration of 0.5 mg/l.

Preparing Stock and Working Solutions

You can either:

- Use recommended concentrations as suggested by suppliers; or
- Make standard strengths as described above; or
- Adapt the solutions according to your working needs.

Example:

You are treating fish stock in 50-l buckets, and you need a treatment dosage of 1 mg/l. You have a 20-ml beaker for dosing the chemical. What strength of solution do you need to give the correct dosage, using one beaker full for every bucket?

The 50 l bucket requires 50 l x 1 mg/l = 50 mg of treatment chemical. Thus you need 50 mg in each 20 ml beaker. Concentration of required solution is therefore 50 mg/20 ml = 50 mg x (1000 ml ÷ 20 ml) = 2 500 mg/l or 2.5 g/I.

Do not try to make concentrations stronger than the solubility limits of the chemical; the supplier will have this information. With dissolving solids, you can visibly check the solubility limit as excess solids will remain suspended in the liquid. In general, however, you should try to make the stock solution quite concentrated, as it will usually keep better and it will be less affected by accidental contamination.

Storing and Dispensing Stock and Working Solutions

Stock and working solutions should be stored in clean bottles, preferably made of dark glass. You may also darken a clearer glass or plastic bottle with black tape. Label the bottle clearly, indicating the kind of chemical, its date of preparation, its strength and notes for its use, for example: malachite green stock solution, prepared 6/6/90, 100 g/I - use 2 ml/100 I to make 2 mg/l solution.

Some chemicals react with other materials, particularly cheap household plastics. Do not use these materials if possible.

Bottles should be well stopped preferably with a glass, rubber or cork top that will not react with the ingredients.

Store the chemicals in a cool place, refrigerated if necessary.

When measuring out the stock or working solution, pour a sample in a clean glass beaker to make sure it has not deteriorated. If acceptable, pour out approximately the amount required, then measure it exactly, using a clean pipette, syringe or measuring cylinder.

Alternatively, draw out the required amount with a clean pipette or syringe, checking to ensure its quality is still acceptable.

As unused working solution might have become contaminated by the measuring equipment or by exposure to the open air, it is preferable not to return it to the stock solution.

Table: Common chemicals to prevent and cure fish diseases.

Chemical	Solid or liquid	Earth ponds	Tanks	Equipment	Live stock	
					Eggs	Juveniles/ adults
Quicklime	S	P	-	-	-	-
Hydraled Lime	S	P	-	-	-	-
Calcium cyanamide	S	P	-	-	-	-
Agro-industrial by-products	S	P	-	-	-	-
Organic poisons	S	p	-	-	-	-
Household bleach	L	-	P	P	-	-
Chlorine bleach liquid	L	-	P	P	-	-
Chlorine bleach powder	S	-	P	P	-	-
Iodophores Wescodyne[R],Romeiod[R],FAM 30[R],etc.	L	-	P	P	P	-
Banzalkonium chlorides Roccal[R],Hyamine[R],etc.	L/S	-	P	P	-	-
Common salt	S	-	-	-	-	P/C
Formalin	L	-	-	-	-	C
Malachite green	S/L	-	-	-	P/C	C
Potassium permanganate	S	-	P	-	-	-
Copper sulphate	S	-	-		-	P/C
Insecticides Bromex[R],Dipterex[R],Neguvon[R],etc.	S	-	-	-	-	C

Some Common Chemicals for Fish Farmers

There are several chemicals which are commonly used by fish farmers to prevent and cure fish diseases.

Limes and calcium cyanamide are particularly useful to control pests in drained ponds.

Agro-industrial by-products such as rice bran, molasses and tobacco dust shavings can be used for similar purposes .

Organic poisons such as rotenone and saponin can control pests in undrained ponds.

Household bleach consists of a weak solution of sodium hypochlorite, which can be used as a good, general disinfectant of non-metallic equipment and working areas. As the active chlorine readily evaporates, a fresh solution should be used within two days.

Chlorine bleach liquid is a stronger solution containing 13 percent active chlorine. It can be diluted or used directly as a strong disinfectant, for example for sterilizing tanks.

Chlorine bleach powder contains 33 percent active chlorine. It makes a very strong disinfectant, and is especially useful for tanks.

lodophores are organic iodine compounds sold as a brownish liquid under various trade names such as Wescodyne, Romeiod, FAM 30 and Buffodine. They are excellent disinfectants, but highly toxic to fish. The active ingredient content varies from one make to another, so check carefully before use. A concentrated solution can be stored for several months in a cool, dark place. The diluted solution remains active for up to a week, until it fades from a brown to straw colour.

Benzalkonlum chlorides are a blend of quaternary ammonium compounds generally sold either as a powder or as a solution under various trade names such as Roccal (10 to 50 percent Al) and Hyamine 3500 (50 percent Al). These are very good disinfectants which can be reused for up to one week.

Common salt as used in the kitchen is usually a cheap and easily available chemical (sodium chloride). In solution, it not only kills several disease organisms but may also have positive effects on the fish by stimulating appetite and increasing mucus secretion, improving resistance to handling. Depending on the species, excess levels may however stress the fish. Thus cyprinids are more susceptible than salmonids.

Formalin is the commercial name for a 35 to 40 percent solution of formaldehyde gas in water. Avoid solutions containing methanol. Formalin is toxic to fish particularly in soft water. As it lowers dissolved oxygen levels, make sure treatment water is well oxygenated. Toxic and irritant for the eyes and lungs, it should be handled very carefully in a well-aerated place. It is very sensitive to light so should be stored in a dark bottle. Always check to see if there is a white deposit at the bottom of the bottle. In such cases, before use, carefully filter out this highly toxic deposit of paraformaldehyde. Keep formalin away from metallic equipment. Remember when determining how much to use that it is considered to be a 100 percent Al chemical.

Malachite green is sold either as a blue-green to green crystalline powder or as a solution of various strengths. You should use the medical-grade, zinc- free quality to avoid heavy fish losses. Do not use in the presence of zinc or iron. Handle carefully and avoid any contact with your skin. If possible, use a trial batch, as quality and toxicity may vary greatly from one batch to another. Not to be used on food fish.

In warm climates, it is safer to use a low dosage mixture of formalin and malachite green, called the Lateux-Meyer mixture. These two chemicals may be mixed and stored together before use.

Potassium permanganate is a violet, crystallized powder. It is a good disinfectant in the absence of organic materials that destroy it. When in solution, it should be kept in a dark bottle.

Copper sulphate is sold as a light-blue powder which readily dissolves in water; blue crystals are also common, but they should be small enough to be easily soluble; it is relatively cheap, but is highly toxic for humans and fish. It should be stored safely and handled properly.

Agricultural insecticides are usually organophosphates and are sold under various trade names such as Bromex, Dipterex, Dylox, Flibol, Masoten and Neguvon. Carefully check the percentage of active ingredients present. They are often very toxic for various other aquatic organisms including zooplankton and for humans and domestic animals. In ponds, they normally decompose within a few days. Toxicity to disease organisms is often reduced when water temperature exceeds 30 °C.

The Toxicity of these Chemicals Vary

Always remember that the toxicity of these chemicals for fish may vary greatly according to water quality. In general toxicity increases:

- As water temperature increases, especially above 25 °C;
- As pH decreases and water becomes acid;
- As total alkalinity decreases, especially below 50 mg/l $CaCO_3$;
- As dissolved oxygen concentration reaches below 5 mg/l.

Check your water quality and look at the chart that follows for the chemical you plan to use. You will learn more about toxicity when considering the choice of disease treatment. Variation of toxicity according to water quality (I toxicity Increases; D toxicity decreases; o no effect).

Chemical	Water t°	pH	Tot. alk.	Organic matter
Formain	I	D	D	-
Copper sulphate	O	D	D	D
Malachite green	O	-	-	-
Potassium permanganate	I	D	-	D
Insecticides	D	-	-	-

Using these Chemicals to Treat Fish

In the dip treatment, fish are kept in a relatively strong solution of the chemical for a very short time, usually less than one minute. Proceed as follows:

- Prepare the chemical solution in a bucket, half-drum or trough;

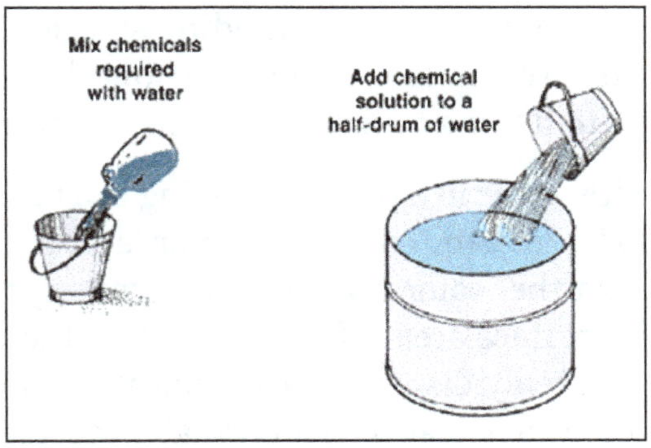

- Place the fish or a batch of fish in a dip net;

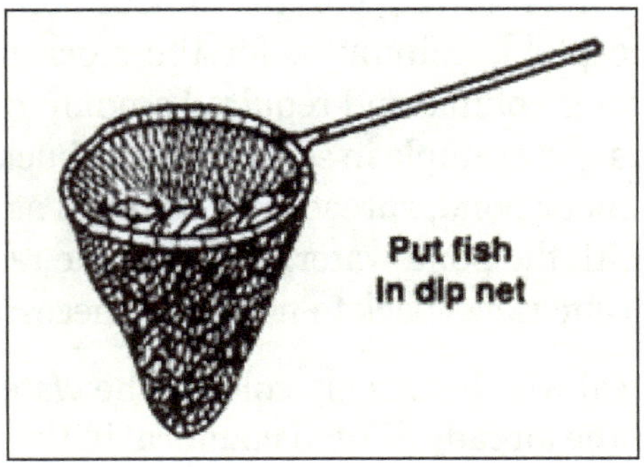

- Dip the fish for the prescribed duration into the solution;

- Immediately after the treatment, replace the fish in well-aerated water.

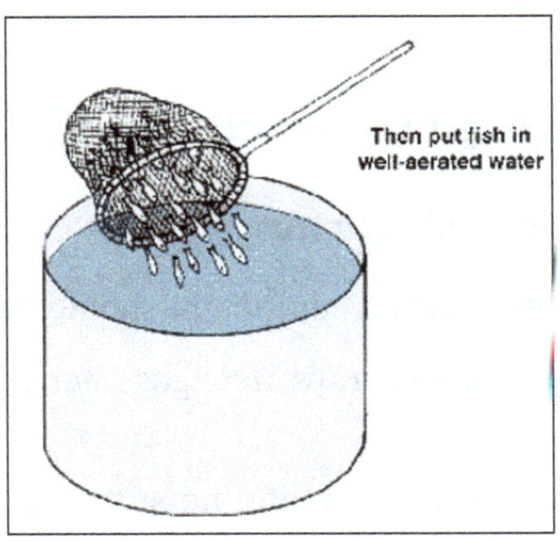

In the bath treatment, fish are kept in a weaker standing solution of the chemical for a longer period of time, which may last from a few minutes to one hour (short bath) in a medium concentration, and up to 24 to 48 h (long bath) in a very low concentration. Proceed as follows:

- For a short bath, mix with water in a plastic watering can the amount of chemical required for the water volume to be treated. Lower the water level in the trough or circular tank by one-third to one-half. Let the water flow in again while spreading the previously mixed chemical over the entire surface area. If necessary, mix the water with a clean broom, an agitator or an aerator to evenly disperse the chemical within the whole water mass. Stop the water inflow once the water reaches its normal level. Treat for the required duration. Then drain two-thirds of the water while starting the water flow again.
- For a long bath, stop the water flowing into the tank or pond. Drain some water and reduce water volume to the acceptable minimum for the stock density and water temperature conditions. Determine water volume and required amount of chemical. Dilute this chemical amount at least 100 times, for example in several plastic buckets before application. Add this diluted solution to the tank or pond, spreading it as much as possible over the entire surface area and mixing it well with the pond water. Treat for the required duration. Then open the water flow and raise the water level back to normal. It necessary, drain again and refill.
- If you are using a chemical which strongly colours the water such as malachite green, you may open the inlet, pour the already diluted chemical in the inflowing water and follow it as it moves across and through the tank or pond. When the colour appears at the outlet, close both inflow and outlet. When the treatment is finished, proceed as before.

In the swab treatment, the concentrated chemical is applied directly to the individual fish, usually with cotton wool or a sponge. This method is usually only used for valuable broodstock fish with specific sores or wounds. The fish should be handled very carefully to avoid the ill effects of additional stress.

There are two other ways to use chemicals but they are more difficult to apply properly and control safely:

- The flush treatment, in which chemicals are added to the inlet of ponds or tanks so that they run through the water volume as a flush; and
- The flow-through treatment, in which a constant supply of chemicals is added to the inflow for a specified time, using a constant-delivery siphon, a container with an adjustable valve or a small electric pump.

Prevention of Diseases through Disinfection

Disinfecting Equipment on Fish Farms

For safe and hygienic practice, the following equipment should be regularly disinfected:

- Hatchery equipment: incubators, trays, troughs, nets, brooms, buckets, tanks, pipes, valves, screens, etc.;
- Handling equipment: containers, handling nets, transport equipment, graders delivery chutes, etc.;

- Boots, waders, protective clothing, vehicle wheels, sampling equipment.

To disinfect this equipment properly proceed as follows:

- Clean it well by brushing and rinsing.
- Apply one of the chemical solutions shown in the following chart, either with a sponge or a brush; whenever possible, immerse the equipment fully into the solution. Preferably use protective gloves.
- Wait for 10 to 15 minutes.
- Rinse thoroughly several times to remove toxicity, before using for fish.

Disinfection of Equipment

Household bleach	• For non-mentallic equipment only • Make a stock solution at 250 ml/l • Use diluted solution = 5 percent stock solution(3 to 4 table-spoons/l)
Benzalkonlum chlorldes	dosage – 200ppm Al • Roccal(25 percent AI): 4 ml/5 l (4 teaspoons/25 l) • Hyamine 3500 (50 percent AI): 2 ml/5 l (2 teaspoons/25 l)

Disinfecting Tanks

Rearing, storage and transport tanks can be disinfected with chemicals as shown in the chart below. Clean the tank thoroughly, especially in corners and drain areas. Apply the recommended solution either by hand with a brush or sponge or using an agricultural sprayer. Protect yourself properly, with gloves, goggles, mask and waders, especially while spraying.

Disinfection of Tanks

Chlorine bleach	for non-metallic tank dosage = 1 000 ppm Al for 20 min or 500 ppm Al for at least 1 h • Chlorine bleach liquid 13 percent Al: 7.5 ml/l (7 500 ppm or about 1 200 ppm Al) for 20 min • Chlorine bleach powder (33 percent Al): 3 ml/l (3 000 ppm or 1 000 ppm Al) for 20 min
Iodophores	dosage = 500 ppm Al for 10 min • FAM 30 2.75 percent Al: 20 ml/l (4 teaspoons/l) • Wescodyne 1.6 percent Al: 30 ml/l (2 tablespoons/l)
Potassium permanganate	dosage: = 1 g/100 l for 15 min

Disinfecting Earthen Ponds

You have already learned earlier:

- How to disinfect drained ponds using quicklime, hydrated lime, calcium cyanamide or, if necessary, agricultural by-products.
- How to disinfect undrained ponds using organic poisons. Remember that this is never as effective as treating drained ponds, which is preferred.

Disinfecting Fish Eggs

Fish eggs can be major sources of transfer of disease from infected broodstock to fry and fingerlings. For this reason, all fish eggs should be thoroughly disinfected before transfer to other facilities. For best results, proceed as follows:

- Freshly prepare a 50 ppm Al solution of an iodophore chemical, for example:
  - Romeiod (0.5 percent Al) 10 ml/I or 2 teaspoons/I;
  - Wescodyne (1.6 percent Al) 10 ml/3 I or 2 teaspoons/3 l;

  You will need plenty of this solution, at least 40 l for 100 000 eggs.
- Adjust the pH of this solution to about 7 with a suitable buffer solution such as sodium bicarbonate (about 100 mg/l).
- Adjust the temperature of this solution to the incubation temperature of the eggs.
- Use this solution to give a short bath of 10 minutes duration to each batch of eggs. When the solution turns yellow, replace it with a fresh one.
- Rinse the eggs well three times with clean water at the incubation temperature.
- Return the eggs to continue their incubation.

Fish eggs can be treated this way either immediately after fertilization or, more frequently, when they reach the eyed stage.

If you run your own hatchery, it is best to treat your eggs regularly against the very common infection caused by the fungus Saprolegnia. There are several treatments, depending on the type of incubator you use.

- If you incubate eggs in a vertical jar proceed as follows:
  - Prepare 1 litre of stock solution containing 500 mg malachite green.
  - Once a day, or more if necessary, stop the water flow in the jar to be treated.
  - Add 2 ml of the stock solution per litre of water present in the jar giving a 1 ppm malachite green solution. In a 7-l jar for example add 2 ml x 7 = 14 ml of stock solution.
  - Mix the chemical well in the jar, taking care not to damage the eggs.

    - Wait for about five minutes.
    - Reopen the water supply, gradually wash the chemical out, and then readjust the flow carefully according to the requirements of the eggs.

- If you incubate eggs in a trough, proceed in a similar way. Stop the water inflow and treat the eggs with a 1 ppm malachite green solution for about five minutes.
- If you cannot treat the eggs within the incubator itself, you should wait until they have developed to the eyed stage. You may then carefully remove the eggs to give them a short bath in a malachite green solution:
    - Either at the concentration of 5 ppm for 20 minutes.
    - At the concentration of 2 ppm for 40 minutes.

Beware: it is always safer to carry out preliminary tests when using a batch of chemicals or a fish species for the first time to determine when and how disinfection can safely be carried out.

Disinfecting Broodstock

Whenever you import new broodstock into your fish farm, it should be disinfected on arrival. Proceed as follows:

- Store the new broodstock in a tank or a small pond, kept separate from the other areas, and with a water supply which does not drain into other rearing units.
- Give a bath of potassium permanganate for one hour, at the dosage of 5 to 10 ppm according to water quality.
- Two days later, give a bath of formalin for at least four hours at the dosage of 10 to 15 ppm according to water quality. Alternatively, you may also use a mixed solution of malachite green and formalin (3.3 g malachite green/I formalin) for one hour at the dosage of 25 ppm.

A simpler but usually less effective way is to use a short bath of 10 minutes in a 2.5 percent salt solution.

After breeding, fish usually become rather susceptible to the development of fungi such as Saprolegnia. You can prevent this from happening through disinfection of the water where you keep your spent broodstock:

- Treat the water every three days for 12 to 15 days.
- Use a bath of malachite green at a dosage of 1 ppm.
- Treat for at least 15 minutes and up to one hour.

for all treatments of fish with chemicals, follow the standard precautions both before and during treatment.

Disinfecting Juvenile Fish before Stocking

When juvenile fish are produced in more intensive systems such as modern hatcheries, it is best to systematically disinfect them before stocking. According to the culture system, this is routinely done for both fry and fingerlings:

- After their harvest from nursery facilities.
- During storage, before their transport.
- During transport, usually within the farm itself.

To disinfect juvenile fish during storage, proceed as follows:

- Prepare the chemical solution in a wide-mouthed plastic container:
    - For advanced fry, use a 2 to 3 percent common salt solution.
    - For fingerlings, use either a 3 to 4 percent common salt solution or a mixed solution containing 2 g/I common saft, 0.2 g/I insecticide and 0.1 ppm malachite green.

prepare a stock solution of malachite green by dissolving 2 g per 10 l of water. Then use 0.5 ml of this stock solution per litre of water to obtain a 0.1 ppm solution.

- Place a pocket of fine-mesh netting within the container.
- Treat successive batches of fish in this netting:
    - If you use common salt only, for a duration of about one minute.
    - If you use the mixed solution, for three to four minutes.
- Then return the fish to clean and well-aerated water for storage until their transport.

To disinfect juvenile fish during transport prepare in advance, for example in a plastic bag or bottle, the. amount of chemical(s) required to treat the volume of water present in the transport container.

If the fish are directly stocked from the container into a pond without further handling, stop the transport three to four minutes before reaching the pond. Pour the chemical solution into the container and dissolve it well into the water. Start the transport again and stock the fish into the pond immediately upon arrival.

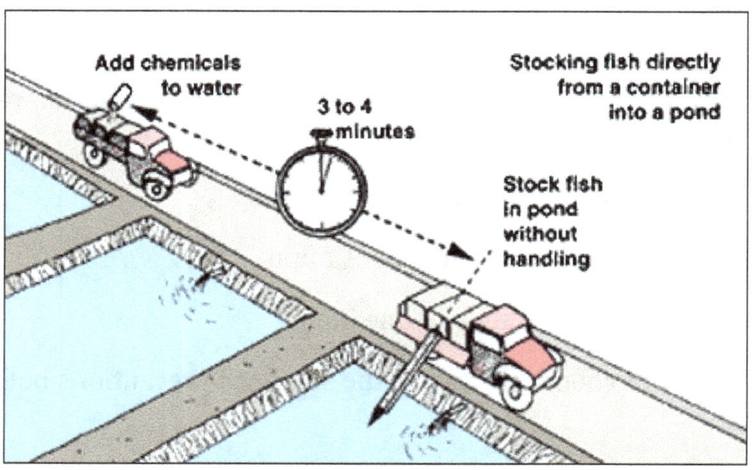

If the fish need to be handled for stocking, pour the chemical(s) into the container three to four minutes before starting to stock the fish into the pond.

Follow the standard precautions both before and during treatment.

Fish Diseases and Possible Cures

Disease Symptoms in Fish

Apart from obvious signs such as dead or dying fish, there are many other symptoms which show that fish are not healthy:

- The behavior of your fish becomes unusual.
- Physical signs are present on the fish.

Observe your fish frequently, for example while feeding them. Learn how they behave and look underwater when healthy to be able to detect as early as possible that something is going wrong. Use the following chart to help you.

Table: Common disease symptoms in fish.

Behavior	Physical signs
• Feeding decreases or stops	• Gaping mouth
• Swimming: weak, lazy, erratic	• Body: open sores, lesions, bloody areas, loss of scales, bloated belly, abnormal coloration
• Floaating in water head up	• Gills: pale, eroded, swollen, bloody or brownish
• Rubbing against hard object	• Fins: folded, eroded
• Flashing / twisting	• Eyes: cloudy or distended
• Darting reoeatedly	• Presence of disease organisms on skin, gills, fins
• Crowding/gathering in shallow water or at water inflow	
• Individual fish isolated from the main group of fish	

Identifying the Cause of Disease

It is not easy to identify in a fish pond why fish show signs of bad health. There are two common situations which you should readily recognize:

- A large part (if not all) of the fish stock show distress or die suddenly, with only some of the above symptoms of disease such as gasping at the surface or gaping mouths: the cause is prior stress (for example rough or poor handling or transport) and/or bad water quality(often low dissolved oxygen) or the presence of a toxic material such as pesticides or other pollution. You have learned earlier what the signs and possible causes are for low oxygen.
- Only a few fish are dead while some others show distress. Usually a few fish die over a period of several weeks and some of the above symptoms are present. The cause is improper

feeding and/or development of some disease organism.Examine the fish closely, observing their swimming behavior and response to feeding. Net a few suspect fish out of the pond and look for visible physical signs of disease, using if possible a magnifying glass. Take samples from skin and gills to be examined under a microscope, either at the farm or in an easily accessible veterinary/medical laboratory.

If in doubt, do not hesitate to call on specialized assistance, for example from a veterinary extension service. Make a careful record of all the circumstances, to provide as much information as possible to assist specialist diagnosis.

Living Organisms Cause Diseases in Fish

There are three major groups of living organisms that may be responsible for fish diseases:

- Viruses,
- Bacteria,
- Parasites.

Viruses are practically invisible organisms. Their detection and identification requires highly specialized laboratory techniques. Control of viral diseases is difficult and requires specialized advice.

Bacteria are minute single-cell organisms (1 to 12 μm), usually living in colonies. Their detection and identification generally also require special laboratory techniques. The treatment of bacterial diseases such as tail or fin rot and skin ulcers requires experienced, specialized advice.

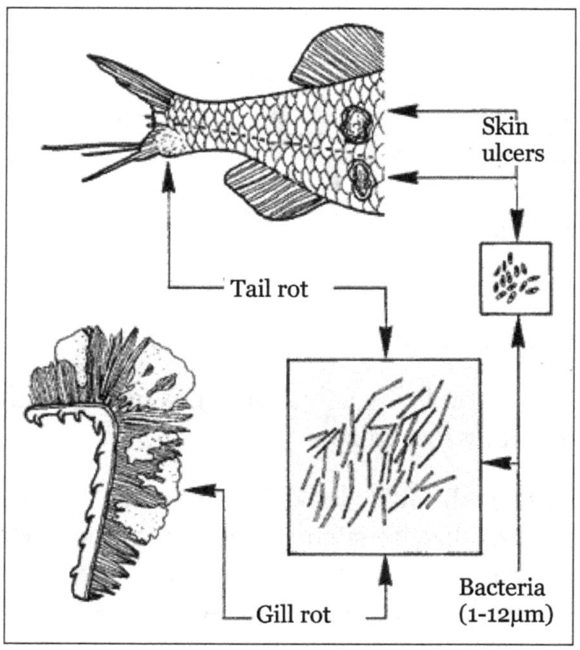

Bacterial diseases.

Parasites are very small to small organisms made up of one or several cells. They develop either inside or outside the body.

- Internal fish parasites are very difficult to control. Although their effects can sometimes be easily identified, detection and identification of the parasites themselves usually requires special skills. Some examples are Myxosoma cerebralis, protozoan cysts in the head of trout causing whirling disease; larval stages (metacercariae) of sucking worms (trematodes), causing black-spot disease and blindness, and tapeworms (Ligula sp.) in the body cavity.

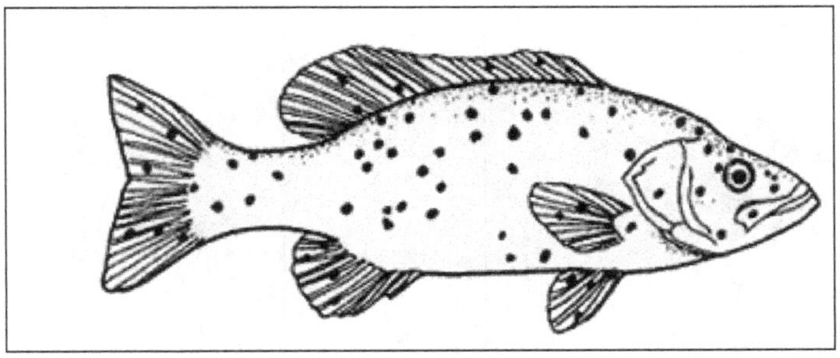

Black-sport disease.

- External fish parasites are much easier to detect and identify. It is usually possible to eliminate them with an appropriate chemical treatment. In the next sections you will learn more about them.

Most Common External Fish Parasites

The most commonly encountered external parasites on a fish farm belong to six different groups.

- Protozoa are very small, single-cell parasites such as:
    - Lchtyobodo sp. (syn. Costia sp.), size 6 to 15 µm.
    - Chilodonella sp., size 30 to 70 µm.
    - Trichodina sp., size 40 to 60 µm.
    - Lchthyophthirius sp. (ich), size 60 to 90 µm.
    - 1 µm = 0.001 mm = one thousandth of one millimetre.
- Flukes (Monogenea) are very small worms attached by hooks, such as Gyrodactylus (body fluke) and Dactylogyrus (gill fluke); size 0.3 to 1 mm.
- Leeches are rather large, segmented worms attached by a sucker on each end, such as Piscicola sp. (3 to 5 cm).
- Copepods often have two elongated egg sacs attached such as Lernaea sp. (anchor worm, 5 to 20 mm) and Ergasilus sp. (size 0.7 to 1.7 mm). Larval stages are similar to the non-parasitic copepods of the zooplankton.
- Fish lice (Crustacea) have a flat, disc-like body covered by a rounded dorsal carapace such as Argulus sp. (6 to 10 mm).

- Water fungi (water moulds) are made of filaments that usually grow into a cotton-like mass or mat such as with Saprolegnia sp.; they can also develop in the gills (Brachiomyces sp.).

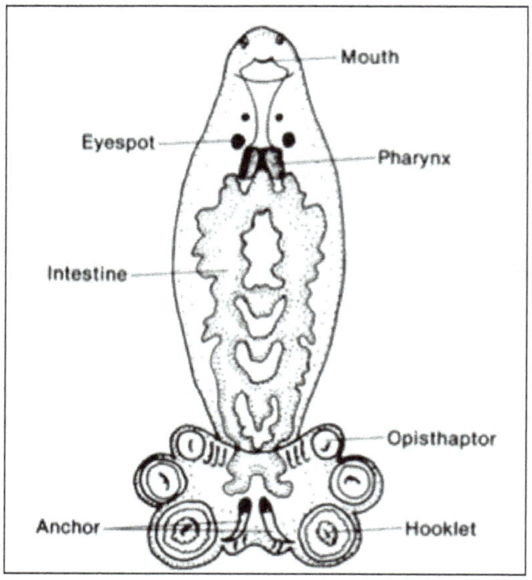

Flukes (Monogenea).

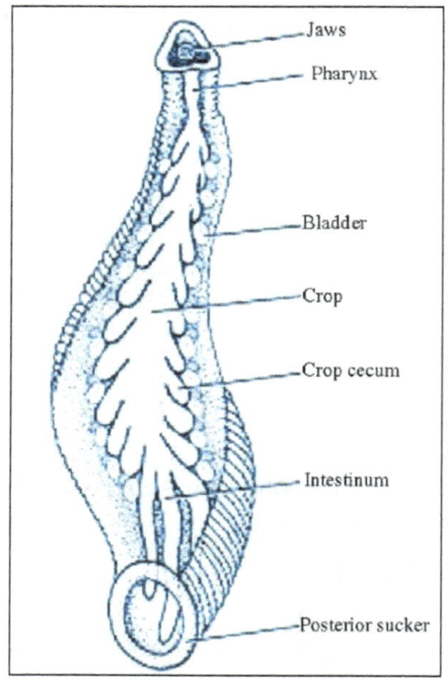

Leeches.

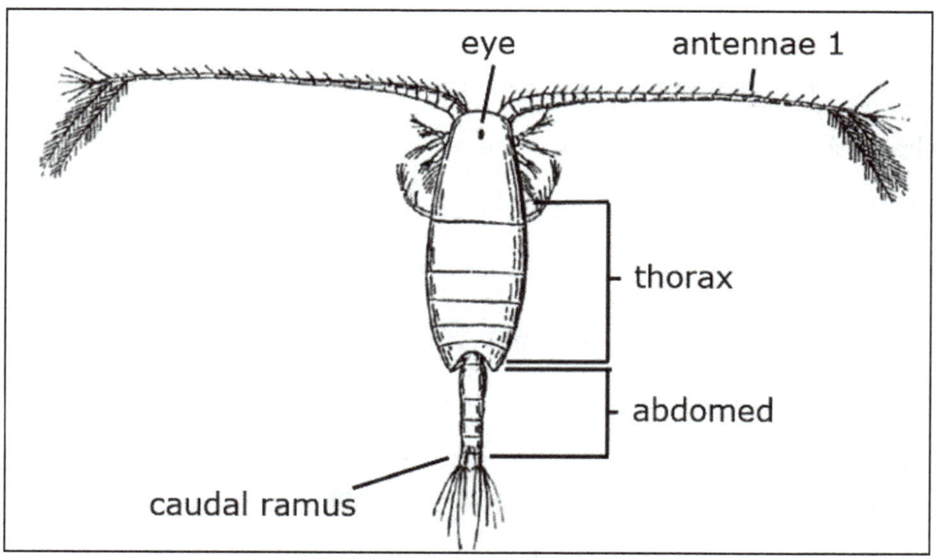

Copepods.

Locating these External Parasites on the Fish

Look for these external parasites on the body skin, the gills, the fins and in the mouth. The base of the fins is a favoured place of attachment.

Some of these parasites are mostly (if not exclusively) found on a particular organ of the fish as shown in the following chart. Some of them also develop on fish eggs, as described earlier.

Location of most common external parasites on fish[1]

Parasite	Body skin	Fins	Gills	Mouth	Eggs	Individual adult Size[2]
Ichthyobodo (Costia)	(x)	(x)	•	30-70 ì m		6-15 ì m
Chilodonella	•	•	•	40-60ì m		
Trichodina	•	•	•	60-90ì m		
Ichthyophthirius	•	•	•			
Gyrodactylus	•	•	(x)			0.3-1 mm
Dactyiogyrus			•			
Leeches	•	(x)	(x)	(x)		3-5 cm
Lernaea	•	•	(x)	•		5-20 mm
Ergasilus		•	•			1.3-1.7 mm
Argulus	•	•		(x)		6-10 mm
Saprolegnia	•	•	(x)	(x)	•	3-30 mm
Brachiomyces			•			

[1] •most frequent (x) occasional

[2] 1ì m =0.001 mm

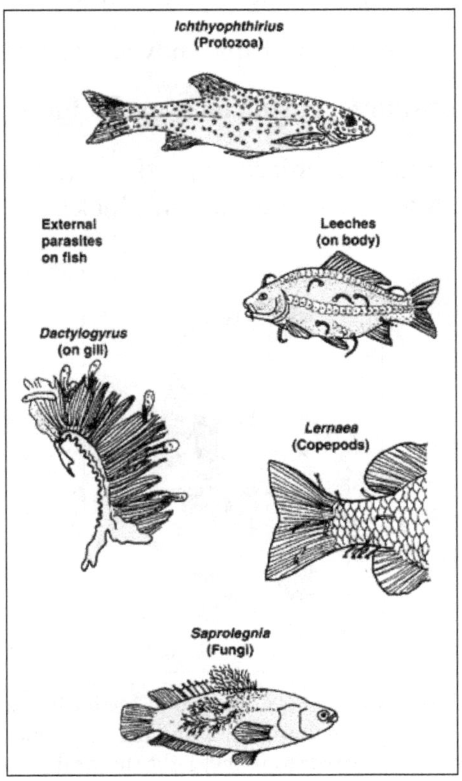

Looking for and Identifying External Fish Parasites

Depending on their size, you will require:

- A magnifying lens, if parasites are at least 1 mm long.
- A microscope, if they are smaller than 1 mm.

You can obtain relatively cheap pocket magnifiers from suppliers selling scientific or optical equipment. More powerful lenses are also available but are more expensive and difficult to use on the farm.

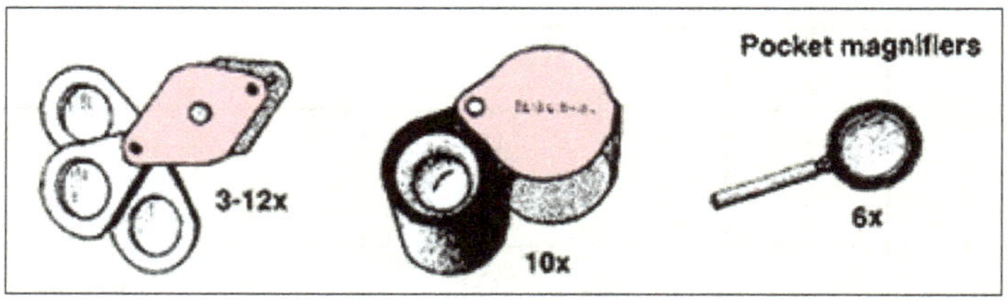

If you have quite a large fish farm, or if you can group together with neighbouring farmers, it would be worthwhile to invest in a good microscope and to learn how to use it properly. You could get assistance in selecting and using a microscope from a local medical or veterinary service, from a farm extension service, or from a local secondary school or college teacher. When buying a microscope, ensure that it has the following items:

- A binocular head;
- Two eyepieces, each with 10x magnification;
- Three objective lenses with 10x, 25x and 40x magnification;
- If possible, a mechanical stage to allow rapid movement of the material during examination;
- If possible, a built-in light source, particularly useful for higher magnification.

In addition you should obtain some glass microscope slides (about 2.5 x 7.5 cm and 1 mm thick) and cover-slips (clear glass, about 2 x 2 cm and 0.2 mm thick).

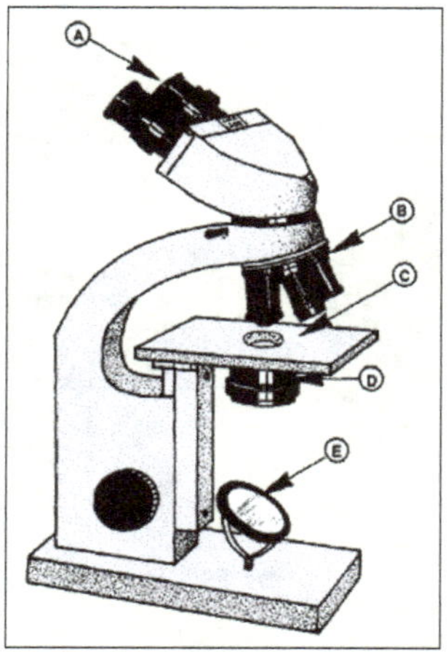

A good microscope for a fish farm: A. Binocular head with two eyepieces (10x) B. Three objective lenses (10x,25x,40x) C. Standard stage (mechanical if possible) D. Condenser E. Mirror to reflect light (natural or artificial).

To look for and identify external parasites with a microscope, proceed as follows:

- For skin/scale samples, collect the mucus (slime) material from the surface of the fish by scraping against the run of the scales.
- Take samples from several skin areas, particularly where the skin is reddened, or where the mucus has a greyish, opaque appearance.
- Place each separate scraping in the centre of a clean glass slide. Add a drop of water and place a cover-slip on top.
- For gill samples, remove a gill arch and cut off the red parts (lamellae). Place some in the centre of a clean glass slide. Add a drop of water and place a cover-slip on top.
- Examine these fresh samples immediately under the microscope under low magnification (objective 10x). Look for signs of movement, which are usually the first clue that parasites are present.
- To identify the parasites, switch your microscope as necessary to medium (objective 25x) or high (objective 40x) magnification.
- Similarly you can mount any small parasite you collect from a fish on a microscope slide in a drop of water, to examine it in more detail.
- To help you identify to which major group a parasite belongs, use the simple key. Start from the top and take the choice that most closely fits the organism you see.

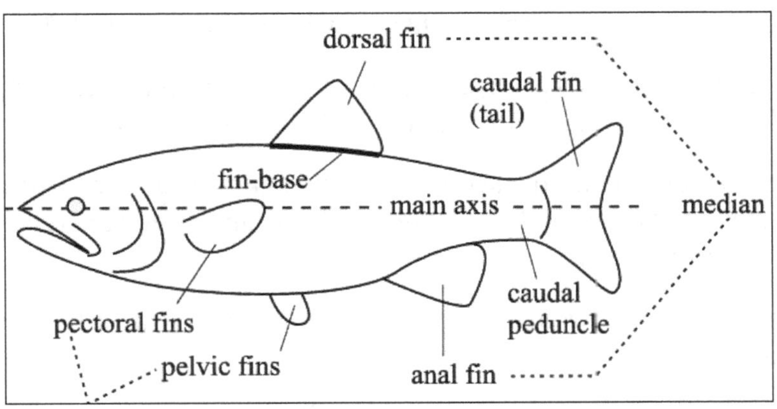

Using Chemicals to Treat your Fish

Once you have accurately identified the agent, you should decide whether or not to treat the fish with chemicals.

- Is a chemical treatment warranted? Are the fish seriously at risk from the parasite, and are there alternate approaches such as improving environmental conditions and removing seriously affected fish?
- Is it economically feasible, given the cost of the chemical, the value of the fish and the risks involved?

- Are the fish still strong enough to survive treatment?
- Is treatment possible in the presence of complicating factors such as an algal bloom, extreme water temperature, low dissolved oxygen or, for example, cloud cover which reduces oxygen production through photosynthesis in the pond?
- Will the chemicals affect any other pond?

Organizing a Treatment

Before initiating any treatment make sure that you have accurately identified the parasite responsible.

Then check water quality thoroughly, in particular the temperature, dissolved oxygen, pH and total alkalinity .Increase water flow, if necessary.

Select the most appropriate curative treatment according to:

- The kind of parasite to be fought;
- Local availability and cost of chemicals;
- Water quality, adapting chemical and dosage to relative toxicity in your own conditions.

Carefully determine the total amount of chemicals required for treatment. Double check on your calculations, preferably with the help of somebody else.

Check the condition of the gills of the fish to be treated. If they have excessive mucus cover and/or are badly formed or chubbed together, the fish may be more susceptible to low oxygen levels.

Always have some additional means of oxygenation standing by, and whenever possible, stop feeding your diseased fish for at least 24 hours immediately before treatment.

Test the selected chemical and its dosage on a small quantity of fish, using exactly the same procedure as for the planned treatment. Observe the effects and after-effects carefully. If necessary modify the treatment accordingly.

Pre-testing becomes even more essential if you plan to use a chemical or to treat a fish species you are not familiar with.

During the treatment, make sure of the following conditions:

- The water temperature is as low as possible. Start the treatment either early in the morning (as long as dissolved oxygen is sufficient) or late in the afternoon.
- The dissolved oxygen content remains adequate. Provide additional oxygen as soon as necessary.
- The chemical is thoroughly mixed with the water mass being treated; dilute the chemical before use according to the kind of treatment. Apply it over the whole water surface area. Use a marker dye such as a few grains of malachite green to check more easily on the mixing of non-coloured chemicals.

- A close watch is kept on the fish being treated. At the first sign of distress such as coming to the water surface, gasping for air or erratic swimming, stop the treatment immediately and provide a good supply of well-aerated, clean water to the fish.
- Complete and accurate records are made, including chemical brand and batch number, dosage, water quality data, fish behavior and losses, etc.

When the treatment is over follow the after-treatment care given here:

- On the next day, check a sample of fish for external parasites. Record your observations carefully.
- Continue to watch the fish for 24 hours. Observe and record possible after-effects of the treatment.
- Repeat the treatment only if it is absolutely necessary and not before two or three days after the previous treatment.

Pretesting a Treatment

Whenever you use an unfamiliar treatment, involving either a new product or a different species, it is better if at all possible to organize preliminary tests to determine the relative toxicity of the chemical under your own conditions.

One simple way of doing this is as follows:

- Prepare a 5- to 10-litre stock solution of the chemical to be tested, according to the concentrations to be tested, as shown in the following chart. For example, if you want to test a new chemical within a range of concentrations 0.25 to 4 ppm active ingredient (Al), prepare a 100 ppm Al stock solution.

Table: Preparing 10-l testing solutions from selected stock solutions.

Stock solution	Testing solution At Al concentration (ppm)	Use stock solution (ml/10 l)
100 ppm Al (0.1 g or ml/l)	0.25 0.50 1 2 4	25 50 100 200 400
1000 ppm Al (1 g or ml/l)	5 10 20 40 80	50 100 200 400 800
5000 ppm Al (5 g or ml/l)	50 100 150 200 250	100 200 300 400 500

- Prepare a series of identical plastic or glass containers of at least 10-l capacity each such as plastic buckets or glass aquaria, avoiding contact with metal:
    - This size is suitable for fish up to 50 g each, slightly more with aeration; for larger fish use correspondingly greater volumes.
    - You will need at least six containers.
    - Two or three sets (12 or 18) of containers would be better; you can then make replicate (double or triple) tests of each concentration.
    - In each set of six, one container should be the control containing no chemical, only clean water; it is used to check that the procedure works properly.
- Add to each of the other five containers the exact volume of stock solution needed to produce the required chemical concentration in a total volume of 10 l.
- Add to each container the exact volume of clean water required to make the total volume up to 10 l. Mix well. The control is filled with 10 l of the same water.
- If possible add extra aeration to the containers using for example an air stone and air pump.
- Add five of the fish to be tested to each container. They should be healthy fish, similar (stock, size, prior treatment) to each other and to the stock about to be treated.
- Carefully observe all fish during the test. Record if and at what time signs of distress are shown. Remove stressed fish promptly to clean and separate holding containers. Observe them until 12 to 24 hours after the normal treatment duration.
- Where possible, pursue the test for the normal duration of the future treatment, and then transfer the remaining batches of fish to containers filled with 10 l of clean fresh water. Observe them for another 12 to 24 hours.
- Record the condition of all the fish at this point. From this, determine the maximum safe dosage of the chemical for the tested fish, the dosage at which all the fish in the batch tolerate the required treatment duration without subsequent ill effects.
- If none of the dosages was tolerated, check the control treatment:
    - If this was not tolerated by all the fish, repeat the trial using better holding conditions (water quality, tank volume, handling, etc.);
    - If the control was tolerated, repeat the test with lower concentrations.
- If all the dosages were tolerated, repeat the trial with higher concentrations.
- The accuracy of this procedure will improve if you use replicates. Run this test with two or three containers for each batch of fish.
- Treat your diseased fish with a dosage at the most equal to the maximum safe dosage.

- If possible, test in a similar way, using infected fish, the effects of a range of safe dosages on the parasite responsible for the disease. Check the survival of parasites after the recommended treatment period. Then treat your fish with the lowest dosage which kills the parasites.

Example:

You wish to test the tolerance of 10- to 20-g juveniles of a new species of fish to a short bath of formalin under your own water quality conditions; proceed as follows.

- For a first series of tests, select a concentration range of 50 to 250 ppm formalin. Accordingly, prepare a 5000 ppm formalin stock solution by placing 50 ml formalin in a plastic bucket and filling it up to the 10-l mark.
- Clean well 12 plastic buckets, providing duplicate batches. In each of them accurately mark with a line the 10 l volume. Clearly number each bucket from 1 to 12 using a permanent marker.
- Add an exact volume of the formalin stock solution to each bucket, as follows:

Bucket no.	Stock Solution (ml)	Tested Concentration (ppm)
1+7	100	50
2+8	200	100
3+9	300	150
4+10	400	200
5+11	500	250
6+12	0	0

- Add clean, well-aerated water to each bucket to fill it up exactly to the 10- l mark. Measure temperature in a few buckets.
- Install air stones and plastic tubing to provide extra aeration to each bucket.
- Add five juvenile fish of the new species to each bucket. Cover the buckets with fine netting to keep the fish from jumping out. Note the exact time at which the tests start, for example 07.40 h.
- Observe the fish for 30 minutes and carefully record distress (D) and removals, which can be summarized as follows:

Time	Bucket no.											
	1	2	3	4	5	6	7	8	9	10	11	12
07.50	-	-	-	-	D2	-	-	-	-	-	D3	-
07.55	-	-	-	D1	D2	-	-	-	-	D2	D2	-
08.05	-	-	-	D2	D1	-	-	-	-	D1	-	-
08.10	-	-	D2	D1	-	-	-	-	D1	D2	-	-

- Move distressed fish in clean water to separate, clearly marked containers, until the end of the trial.
- Then, exchange solutions in each bucket with clean fresh water; move all tested fish back into their respective containers. Observe the fish for about 12 hours and note that some of the fish in buckets 4, 5, 10 and 11 are still not fully recovered by the end of the test at 20.00 h.
- Conclusion: the maximum safe dosage for a 30-minute bath is 100 ppm formalin for 10- to 20 g juveniles of the new fish species, in the conditions of the test, in particular water temperature ranging from 21°C to 22°C.
- Before treating all your fish infected with external parasites (for example Trichodina protozoa), run a similar test with a few diseased fish in formalin solutions ranging from 50 ppm to 100 ppm (for example 50, 60, 70, 80, 90 and 100 ppm solutions made from the 5 000 ppm stock solution). Determine which is the lowest safe concentration that eliminates the parasites in a 30-minute bath.

Choosing Chemicals to Control External Parasites

All chemicals are not fully effective to control all external fish parasites or fungi. After identifying the parasite you must control, you should select the most appropriate chemical, according to the following chart.

If the best chemical is not available, a treatment may be attempted with some other one, as shown below.

Table: Chemicals most effective to control particular parasites or fungi.

Parasite	Formalin or Formalin+malachite green	Organophosphate Insecticide	Malachite green
Protozoa	•		
Body flukes	•		
Gill flukes		•	
Leeches		•	
Fish lice		•	
Copepods		•	
Saprolegnia			•

Choosing Treatment

Examples of treatments which have been successfully applied to farmed fish in various countries to control external parasites and fungi are summarized in above table. They involve the use of various chemicals:

- For short baths given in tanks and troughs.
- For long baths given in small earthen ponds.

- For dips given to fish batches in various containers, using the method described earlier.

Always remember that the chemical dosages and treatment durations given in are guidelines only. You have already learned earlier how the toxicity of chemicals for fish varies with water quality. But it may also change from one batch of chemicals to another:

- With the fish species.
- With the age of the fish.

Therefore, if you are not familiar with a batch of chemicals or a kind of fish, make a preliminary test before treating your fish stock. In all cases, closely follow the procedures suggested earlier.

In addition to what you have learned previously about these chemicals, the following points are particularly important.

- Formalin can damage the gills of fish. Its toxicity greatly increases as the water temperature increases. In warm water you should prefer lower dosages of a mixture of formalin and malachite green. Make sure the dissolved oxygen is always above 5 mg/l during treatment, as formalin removes oxygen from water. Salmonids are less tolerant than other fish.
- Formalin and malachite green mixture can be prepared in two ways:
    - Quality A = add 3.68 g malachite green per litre formalin.
    - Quality B = add 3.30 g malachite green per litre formalin.
    - Do not use if water temperature is above 28 °C and/or if dissolved oxygen is below 5 mg/l.
- Malachite green quality often varies from one batch to another. Be sure you buy the zinc-free grade. Salmonids are more tolerant to this chemical than other fish. Chinese carp have little tolerance. There is a tendency to ban the use of this chemical in certain countries. Do not use on food fish.
- Organophosphates lose their effectiveness on parasites at water temperatures above 30 °C. Salmonids are much more susceptible to them than are cyprinids or tilapias. There are two kinds of organophosphates:
    - In quality C, the active ingredient is metrifonate (Bayer) as in commercial brands Chlorofas, Dipterex, Dylox, Flibol, Masoten, Neguvon, Trichlorphon and Tugon;
    - In quality D, the active ingredient is dichlofenthion as in commercial brands Bromex, Dibrom and Naled (these are more toxic to fish but also usually more effective to control parasites at smaller dosages).
- Common salt: tolerance varies with species and age:
    - Salmonids are more tolerant than cyprinids.
    - Adults are more tolerant than juveniles.

- Copper sulphate toxicity greatly varies with water alkalinity. Do not use it if total alkalinity is less than 50 mg/l $CaCo_3$ .Remember that heavy rains may suddenly decrease the alkalinity of stream and pond waters. To treat fish in an earthen pond using a long bath, apply about 0.75 ppm copper sulphate for every 100 ppm $CaCo_3$ total alkalinity. Do not use if total alkalinity is above 400 ppm $CaCo_3$. Watch the effect of copper sulphate on phytoplankton and aquatic vegetation. Oxygen levels may be reduced.

References

- Aquaculture, office of. “basic questions about aquaculture :: office of aquaculture”. Www.nmfs.noaa.gov. Retrieved 2016-06-09
- Fish-farming, energy-government-and-defense-magazines, environment: encyclopedia.com, Retrieved 26 July, 2019
- Weaver, d e (2006). “design and operations of fine media fluidized bed biofilters for meeting oligotrophic water requirements”. Aquacultural engineering. 34 (3): 303–310. Doi:10.1016/j.aquaeng.2005.07.004
- Advantages-of-fish-farming-business, agripedia: krishijagran.com, Retrieved 21 May, 2019
- Torrissen, ole; et al. (2011). “atlantic salmon (salmo salar): the ‘super-chicken’ of the sea?”. Reviews in fisheries science. 19 (3): 257–278. Doi:10.1080/10641262.2011.597890
- Relating to labeling and identification of genetically modified fish and fish products”. Alaskan senate bill no. 25 of 19 may 2005. Retrieved 2 december 2017

Fishing Techniques

CHAPTER 3

Various kinds of techniques are used in fishing. These include remote control fishing, gillnetting, longline fishing, seine fishing, trawling, electric pulse fishing, electrofishing, etc. This chapter closely examines these types of fishing techniques to provide an extensive understanding of the subject.

TROLLING

Trolling is method of fishing in which a lure or a bait is pulled behind a boat at varying speeds and depths according to the nature, habitat, and size of the fish being sought. Trolling is practiced in both freshwater and salt water and with all kinds of craft; power boats that carry varied tackle and big-game gear are usually used at sea but may also be used on inland waters, where legal. Trolling permits the fisherman to cover a wide area, making it the method of choice among sport anglers in pursuit of such highly mobile species as walleye and muskellunge (muskie) in large freshwater lakes and rivers. Trolling for sport is usually done at relatively low speeds using strong rods equipped with stout lines and heavy reels. The fisherman sits facing the stern, sometimes in a specially equipped fighting chair. Troll lines are also used by trawler fishermen for commercial catches, principally salmon and tuna.

In trolling, lines bearing baited hooks or lures are towed through the water by a slow moving vessel. As with any hook and line fishing method, fish are caught when they snap at baited hooks which then become embedded in the fish's mouth or elsewhere.

Hooking is stressful to fish, as discussed in rod and line fishing. Troll-caught salmon have been shown to be severely exhausted when landed. Hooking can also result in severe injury.

As with all forms of hook and line fishing, sometimes fish are gaffed (i.e. impaled on a hook) to bring them abroad and sometimes live fish are used as bait. These practices are likely to cause considerable additional suffering.

Environmental Impact

As with rod and line fishing, conservation groups consider trolling to have low levels of bycatch relative to other major fishing methods. Bycatch fish can be released quickly.

Rod and line fishing, levels of hooking injury, and the survival chances of released bycatch fish, are affected by the type/size of the hook/bait and the method by which the hook is removed.

One study looked at death rates for Chinook salmon that had been caught off Alaska by trolling, and subsequently released as bycatch. It found that many troll-caught fish become hooked in locations other than the mouth. The study found that this increases the likelihood of fatal wounding, especially if the gills are damaged. For fish hooked through the gills, 85% died. Nearly a quarter of the fish, 23%, were hooked through the eye, of which 21% of them died within 6 days.

The same study found that smaller fish may cope less well with being hooked than those of legal capture size. An earlier study of the same fishery found that the hooking location seemed to be affected by the type of lure.

Measures to Reduce Suffering in Trolling

The following measures, combined with humane slaughter immediately the fish is landed, would improve the welfare of fish caught in trolling.

Reducing the suffering in trolling:

Reduce Suffering of Bait Fish

- Avoid the use of live bait fish.
- Avoid the use of bait fish generally (use artificial baits or off cuts instead).

Keep the Duration of Capture Short

- Monitor gear and land fish immediately they become hooked (fair-fish limits capture duration to 5 minutes for fish caught by hook).

Reduce the Numbers of Bycatch Animals

- Use hooks and baits that reduce bycatch.

Reduce Stress and injury during Capture

- Use hooks than cause less injury e.g. circle hooks, barbless circle hooks.

Reduce Stress and Injury during Landing

- Handle fish carefully when landing prior to humane slaughter (or release as bycatch).
- Minimise time spent out of water.
- Remove hooks after fish is humanely slaughtered or stunned, rather than before.
- Carefully remove hooks from fish to be released.
- Avoid gaffing fish.

Keys to Trolling

The secret to consistently successful trolling isn't using special equipment or finding a hot place. It's presenting your lure most effectively. If more people really understood this, fewer of them would be wandering around doing the wrong thing and unintentionally depending on luck.

Like any other method of fishing, trolling demands figuring out where and how to make the best presentation. No matter what species you seek, the following factors are common, and critical, to effective trolling.

Depth Control for Trolling Lures

The most important of these factors is knowing and controlling where your lure is. You can't catch fish consistently unless your lure is in the right place. Expecting any species to come up or go down more than a short distance is unrealistic; most will only come up a little, and few go down. In trolling, a lure goes by a fish pretty quickly; the fish doesn't have a lot of time to think about striking. If the lure isn't close to the level of the fish, the chance of seeing it let alone striking it is far less.

Speed

Most anglers don't fully appreciate the vital role speed plays in trolling, which may explain why trolling is a hit-or-miss proposition for so many. Speed relates to the behavior of the fish you seek; the type of lure or bait being used; and to boat maneuvering techniques.

No matter what kind of fish you troll for, or what tackle and type of boat you use, you'll get more out of your lures by paying close attention to the speed at which they're working. The correct speed is the one that gets the right action out of your lures and is correct for the fish you seek.

Lure Action

The pulling of a lure by the boat is akin to retrieving a lure that is cast, except that more variables can affect lure action. Some of these are quite obvious and can be controlled or countered, including current, waves, wind velocity and direction, type and weight of boat, and power of the engine.

Some are more subtle, however, and need to be in the minds of every troller. For example, not all lures work well at the same speed. Some lures are incompatible with others because they require different speeds to work best. Be careful when mixing lure types while trolling. Some lures only work well at slow speeds and may run awry when trolled quickly, thus having almost no chance of catching a fish.

Side and Back Placement

Placing lures at a specific depth is a major element in trolling, but so is considering how far they should be behind the boat and to the side of the boat. Some fish — usually in clear water, in shallow water, or near the surface — are extremely wary, which greatly affects how and where you set trolled lures. When open-water fish are shallow, you need to get lures horizontally distant from a boat, since the fish will move away from the noise of the moving boat. This is where sideplaners and planer boards come into play because they spread trolling lines out. Directional diving planers can do the same thing; when adjusted to run off to the side, they take a lure down and outward.

The distance behind the boat, which is called setback, is a separate issue. Setback distance depends on the type of fish, water clarity, boat traffic, and lures being used, as well as the method of fishing.

Boat Control

Where a trolled lure is at any given time is largely dependent on how you maneuver the boat that pulls it, and how you counteract elements that would affect your trolling, especially wind and current.

One of the greatest mistakes made by trollers is to fish at the same boat speed when heading into the wind as when moving with the wind. Another is following the same straight path endlessly. It is no coincidence that many fish are caught when boaters speed up or slow down and when they make turns.

REMOTE CONTROL FISHING

Remote control fishing is a fishing technique accomplished by using a remote control boat. The remote control fishing boats are almost exclusively battery operated, since batteries provide several benefits when compared with glow engines: less noise during operation, consume no power during idle time, requires no oxygen, easier to maintain, finer power and speed control, etc.

Control Fishing Boats

Remote control fishing boats are either modified RC boats or custom build RC fishing boats. Theoretically, as long as a fishing line can be attached to the remote control boat, any remote control boat can be used. But, good electric RC fishing boat should be:

- Water tight up to a certain depth - if pulled down by larger fish, it should stay operational after being submerged,
- Visible from long distances - if fishermen lose the RC boat out of sight, it is very hard to bring it back again,
- Have enclosed propellers or small jets - fishing lines, sea-grass and various debris can get entangled around unprotected propellers and can prevent the RC boat from operating normally,
- Rudder should be omitted or should be enclosed with propellers - if two electric engines are used, rudder can be omitted and boat steering is done using electric motors,
- Have low center of gravity - boat is stable and can handle bad weather, waves and wind,
- Should be mono-hull - mono-hull RC boats with low center of gravity always position them properly (even after being fully upside down) and, if they are water tight, can resume

operation right away. Catamaran RC fishing boats are very stable boats, but when turned upside-down due to wind, waves, fish etc., they are unable to return to proper position on their own. Regardless of hull type, hulls should be strong, durable and water tight and should be able to withstand hits and impacts with boats, debris, rocks etc.

- Batteries should be swapped easily, with minimum tools required. NiMh batteries are common in most RC boats, with NiCd batteries being slowly phased-out due to environmental concerns, battery memory effect, lower capacity and generally being inferior to NiMh batteries. High-end models of RC boats are mostly powered with strong, lightweight and expensive lithium polymer batteries.
- Electric motors should be brushless electric motors. Brushless motors are more expensive and require more complex motor controllers, but they require less maintenance and are more reliable than brushed electric motors,
- Boat should be able to achieving various (trolling) speeds, not just to have simple motors 'off and on'. etc.

Fishing Methods

There are several fishing methods that are usually employed:

- A section of fishing line and baited hook are permanently attached to a remote control boat. Using remote control, bait is positioned on desired location and fishermen either wait for a fish to strike or use RC boat for trolling. The limitation of this method is if a big fish is hooked, there is a risk of the fish pulling the remote control boat underwater.
- Attach a line release mechanism to the RC boat and then attach the line/hook and (instead of casting) drive the line out with the RC boat. When a fish strikes, the line disconnects, and the fisherman reels in the fish with a regular fishing pole. This allows the fisherman to catch any size fish.
- Similar to the previous method; RC fishing boat is equipped with line release mechanism and when on desired position, line is released by fishing rod strike or by remote command. After fishing line is released, RC boat is retrieved using remote control or even automatically using some sort of homing system (ultra sound, infra red, GPS, etc.).

RC fishing methods are limited only by imagination of RC fishing boat users.

Line Release Mechanism

Line release mechanism can be attached to any suitable RC boat. Instead of using fast electric RC boats, slower, but stronger and more durable RC boats should be used. Line release mechanism is mounted on the stern of the boat, so that the fishing line stays clear of propellers, rudders, antenna and any other protrusions.

Also, line release mechanism can be improvised using various methods, mostly using ordinary plastic clothespin requiring some testing before such system can be actually used.

TRAWLING

Trawling is a method of fishing that involves pulling a fishing net through the water behind one or more boats. The net that is used for trawling is called a trawl.

The boats that are used for trawling are called trawlers or draggers. Trawlers vary in size from small open boats with as little as 30 hp (22 kW) engines to large factory trawlers with over 10,000 hp (7.5 MW). Trawling can be carried out by one trawler or by two trawlers fishing cooperatively (pair trawling).

Trawling can be contrasted with trolling, where baited fishing lines instead of trawls are drawn through the water. Trolling is used both for recreational and commercial fishing whereas trawling is used mainly for commercial fishing. Trawling is also commonly used as a scientific sampling, or survey, method.

Bottom vs. Midwater Trawling

Trawling can be divided into bottom trawling and midwater trawling, depending on how high the trawl (net) is in the water column. Bottom trawling is towing the trawl along (benthic trawling) or close to (demersal trawling) the sea floor. Midwater trawling is towing the trawl through free water above the bottom of the ocean or benthic zone.

Midwater trawling is also known as pelagic trawling. Midwater trawling catches pelagic fish such as anchovies, shrimp, tuna and mackerel, whereas bottom trawling targets both bottom-living fish (groundfish) and semi-pelagic fish such as cod, halibut and rockfish.

The gear itself can vary a great deal. Pelagic trawls are typically much larger than bottom trawls, with very large mesh openings in the net, little or no ground gear, and little or no chaffing gear. Additionally, pelagic trawl doors have different shapes than bottom trawl doors, although doors that can be used with both nets do exist.

Net Structure

Bottom trawling.

When two boats are used (pair trawling), the horizontal spread of the net is provided by the boats, with one or in the case of pelagic trawling two warps attached to each boat. However, single-boat trawling is more common. Here, the horizontal spread of the net is provided by trawl doors (also known as "otter boards"). Trawl doors are available in various sizes and shapes and may be specialized to keep in contact with the sea bottom (bottom trawling) or to remain elevated in the water. In all cases, doors essentially act as wings, using a hydrodynamic shape to provide horizontal spread. As with all wings, the towing vessel must go at a certain speed for the doors to remain standing and functional. This speed varies, but is generally in the range of 2.5–4.0 knots.

The vertical opening of a trawl net is created using flotation on the upper edge ("floatline") and weight on the lower edge ("footrope") of the net mouth. The configuration of the footrope varies based on the expected bottom shape. The more uneven the bottom, the more robust the footrope configuration must be to prevent net damage. This is used to catch shrimp, shellfish, cod, scallops and many others. Trawls are funnel-shaped nets that have a closed-off tail where the fish are collected and is open on the top end as the mouth.

Trawl nets can also be modified, such as changing mesh size, to help with marine research of ocean bottoms.

Environmental Effects

Nets for trawling in surface waters and for trawling in deep water and over the bottom.

Although trawling today is heavily regulated in some nations, it remains the target of many protests by environmentalists. Environmental concerns related to trawling refer to two areas: the lack of selectivity and the physical damage which the trawl does to the seabed.

Selectivity

Since the practice of trawling started (around the 15th century), there have been concerns over trawling's lack of selectivity. Trawls may be non-selective, sweeping both marketable and undesirable fish and fish of both legal and illegal size. Any part of the catch which cannot be used is considered by-catch, some of which is killed accidentally by the trawling process. By-catch commonly includes valued species such as dolphins, sea turtles, and sharks, and may also include sublegal or immature individuals of the targeted species.

Many studies have documented large volumes of by-catch that are discarded. For example, researchers conducting a three-year study in the Clarence River found that an estimated 177 tons of by-catch (including 77 different species) were discarded each year.

Size selectivity is controlled by the mesh size of the "cod-end"—the part of the trawl where fish are retained. Fishermen complain that mesh sizes which allow undersized fish to escape also allow some legally catchable fish to escape. There are a number of "fixes", such as tying a rope around the "cod-end" to prevent the mesh from opening fully, which have been developed to work around technical regulation of size selectivity. One problem is when the mesh gets pulled into narrow diamond shapes (rhombuses) instead of squares.

The capture of undesirable species is a recognized problem with all fishing methods and unites environmentalists, who do not want to see fish killed needlessly, and fishermen, who do not want to waste their time sorting marketable fish from their catch. A number of methods to minimize this have been developed for use in trawling. Bycatch reduction grids or square mesh panels of net can be fitted to parts of the trawl, allowing certain species to escape while retaining others.

Studies have suggested that shrimp trawling is responsible for the highest rate of by-catch.

Environmental Damage

Setting a trawl.

Trawling is controversial because of its environmental impacts. Because bottom trawling involves towing heavy fishing gear over the seabed, it can cause large-scale destruction on the ocean bottom, including coral shattering, damage to habitats and removal of seaweed. The primary sources of impact are the doors, which can weigh several tonnes and create furrows if dragged along the bottom, and the footrope configuration, which usually remains in contact with the bottom across the entire lower edge of the net. Depending on the configuration, the footrope may turn over large rocks or boulders, possibly dragging them along with the net, disturb or damage sessile organisms or rework and re-suspend bottom sediments. These impacts result in decreases in species diversity and ecological changes towards more opportunistic organisms. The destruction has been likened to clear-cutting in forests.

The primary dispute over trawling concerns the magnitude and duration of these impacts. Opponents argue that they are widespread, intense and long-lasting. Defenders maintain that impact is mostly limited and of low intensity compared to natural events. However, most areas with significant natural sea bottom disturbance events are in relatively shallow water. In mid to deep waters, bottoms trawlers are the only significant area-wide events.

Bottom trawling on soft bottoms also stirs up bottom sediments and loading suspended solids into the water column. One bottom trawler can put more than 10 times the amount of suspended solids pollution per hour into the water column than all the suspended solids pollution from all the sewerage, industrial, river and dredge disposal operations in Southern California combined. These turbidity plumes can be seen on Google Earth in areas where they have high resolution offshore photos. When the turbidity plumes from bottom trawlers are below a thermocline, the surface may not be impacted, but less visible impacts can still occur, such as persistent organic pollutant transfer into the pelagic food chain.

As a result of these processes, a vast array of species are threatened around the world. In particular, trawling can directly kill coral reefs by breaking them up and burying them in sediments. In addition, trawling can kill corals indirectly by wounding coral tissue, leaving the reefs vulnerable to infection. The net effect of fishing practices on global coral reef populations is suggested by many scientists to be alarmingly high.

Midwater (pelagic) trawling is a much “cleaner” method of fishing, in that the catch usually consists of just one species and does not physically damage the sea bottom. However, environmental groups have raised concerns that this fishing practice may be responsible for significant volumes of by-catch, particularly cetaceans (dolphins, porpoises, and whales).

Regulation

In light of the environmental concerns surrounding trawling, many governments have debated policies that would regulate the practice.

Other uses of the Word “trawl”

The noun “trawl” has many possibly confusing meanings in commercial fisheries. For example, two or more lobster pots that are fished together may be referred to as a trawl. In some older usages “trawling” meant “long-line fishing”; that usage occurs in Rudyard Kipling’s book *Captains*

Courageous. (This use is perhaps confused with trolling, where a baited line is trailed behind a boat. Troll also has several meanings.)

The word "trawling" has come to be used in a number of non-fishing contexts, usually meaning indiscriminate collection with the intent of picking out the useful bits. For instance, in law enforcement it may refer to collecting large volumes of telephone call records hoping to find calls made by suspects. It also occurs frequently in reference to research methods, where it means searching through written sources for relevant information.

Bottom Trawling

Bottom trawling is trawling (towing a trawl, which is a fishing net) along the sea floor. It is also referred to as "dragging". The scientific community divides bottom trawling into benthic trawling and demersal trawling. Benthic trawling is towing a net at the very bottom of the ocean and demersal trawling is towing a net just above the benthic zone.

Bottom trawling can be contrasted with midwater trawling (also known as pelagic trawling), where a net is towed higher in the water column. Midwater trawling catches pelagic fish such as anchovies, and mackerel, whereas bottom trawling targets both bottom-living fish (groundfish) and semi-pelagic species such as cod, squid, shrimp, and rockfish.

Trawling is done by a trawler, which can be a small open boat with only 30 hp (22 kW) or a large factory trawler with 10,000 hp (7,500 kW). Bottom trawling can be carried out by one trawler or by two trawlers fishing cooperatively (pair trawling).

Global catch from bottom trawling has been estimated at over 30 million tonnes per year, an amount larger than any other fishing method. Concerns about the environmental impacts of bottom trawling have led to changes in gear design, such as the addition of turtle excluder devices to reduce bycatch, and limitations on locations where bottom trawling is allowed, such as marine protected areas.

A bottom trawler at Viveiro in Spain.

An early reference to fishery conservation measures comes from a complaint about a form of trawling dating from the 14th century, during the reign of Edward III. A petition was presented to

Parliament in 1376 calling for the prohibition of a "subtlety contrived instrument called the *wondyrchoum*". This was an early beam trawl with a wooden beam, and consisted of a net 6 m (18 ft) long and 3 m (10 ft) wide,

> "of so small a mesh, no manner of fish, however small, entering within it can pass out and is compelled to remain therein and be taken by means of which instrument the fishermen aforesaid take so great abundance of small fish aforesaid, that they know not what to do with them, but feed and fatten the pigs with them, to the great damage of the whole commons of the kingdom, and the destruction of the fisheries in like places, for which they pray remedy."
>
> "three fathom long and ten mens' feet wide, and that it had a beam ten feet long, at the end of which were two frames formed like a colerake, that a leaded rope weighted with a great many stones was fixed on the lower part of the net between the two frames, and that another rope was fixed with nails on the upper part of the beam, so that the fish entering the space between the beam and the lower net were caught. The net had maskes of the length and breadth of two men's thumbs".
>
> –The Wondyrchoum

The response from the Crown was to "let Commission be made by qualified persons to inquire and certify on the truth of this allegation, and thereon let right be done in the Court of Chancery". Thus, already back in the Middle Ages, basic arguments about three of the most sensitive current issues surrounding trawling - the effect of trawling on the wider environment, the use of small mesh size, and of industrial fishing for animal feed - were already being raised.

Until the late 18th century sailing vessels were only capable of towing small trawls. However, in the closing years of that century a type of vessel emerged that was capable of towing a large trawl, in deeper waters. The development of this type of craft, the sailing trawler, is credited to the fishermen of Brixham in Devon. The new method proved to be far more efficient than traditional long-lining. At first its use was confined to the western half of the English Channel, but as the Brixham men extended their range to the North Sea and Irish Sea it became the norm there too.

By the end of the 19th century there were more than 3,000 sailing trawlers in commission in UK waters and the practice had spread to neighbouring European countries. Despite the availability of steam, trawling under sail continued to be economically efficient, and sailing trawlers continued to be built until the middle of the 1920s. Some were still operating in UK waters until the outbreak of World War II, and in Scandinavia and the Faroe Islands until the 1950s.

English commissions in the 19th century determined that there should be no limitation on trawling. They believed that bottom trawling, like tilling of land, actually increased production. As evidence, they noted that a second trawler would often follow a first trawler, and that the second trawler would often harvest even more fish than the first. The reason for this peculiarity is that the destruction caused by the first trawl resulted in many dead and dying organisms, which temporarily attracted a large number of additional species to feed on this moribund mass.

Bottom trawling does not only have a long tradition in European waters. It was also recognized in 1704 during the Edo era in Japan as a common fishing method. A slightly different approach was developed where the "Utase Ami" or "trawled bottom net" was deployed from a sideways sailing boat.

Bottom trawling has been widely implicated in the population collapse of a variety of fish species, locally and worldwide, including orange roughy, barndoor skate, shark, and many others.

Fishing Gear

The design requirements of a bottom trawl are relatively simple, a mechanism for keeping the mouth of the net open in horizontal and vertical dimensions, a "body" of net which guides fish inwards, and a "cod-end" of a suitable mesh size, where the fish are collected. The size and design of net used is determined by the species being targeted, the engine power and design of the fishing vessel and locally enforced regulations.

Beam Trawling

A beam trawl.

The simplest method of bottom trawling, the mouth of the net is held open by a solid metal beam, attached to two "shoes", which are solid metal plates, welded to the ends of the beam, which slide over and disturb the seabed. This method is mainly used on smaller vessels, fishing for flatfish or prawns, relatively close inshore.

Otter Trawling

Otter trawl.

Otter trawling derives its name from the large rectangular otter boards which are used to keep the mouth of the trawl net open. Otter boards are made of timber or steel and are positioned in such a way that the hydrodynamic forces, acting on them when the net is towed along the seabed, push them outwards and prevent the mouth of the net from closing. They also act like a plough, digging up to 15 cm into the seabed, creating a turbid cloud, and scaring fish towards the net mouth. The net is held open vertically on an otter trawl by floats and/or kites attached to the "headline" (the rope which runs along the upper mouth of the net), and weighted "bobbins" attached to the "foot rope" (the rope which runs along the lower mouth of the net). These bobbins vary in their design depending on the roughness of the sea bed which is being fished, varying from small rubber discs for very smooth, sandy ground, to large metal balls, up to 0.5 m in diameter, for very rough ground. These bobbins can also be designed to lift the net off the seabed when they hit an obstacle. These are known as "rock-hopper" gear.

Body of the Trawl

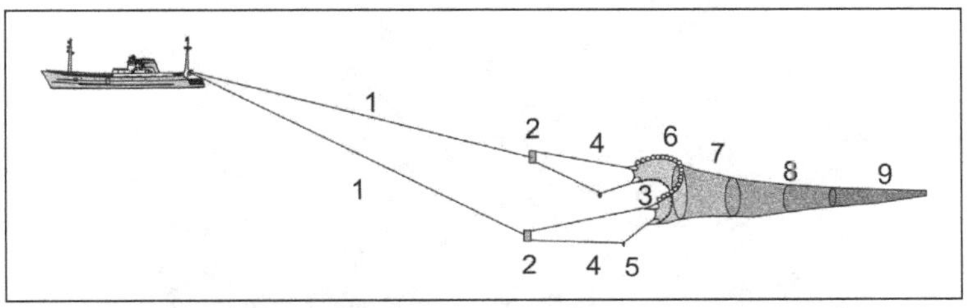

Pelagic (midwater) otter trawl. 1: trawl warp, 2: otter boards, 3: longline chains, 4 hunter, 5: weights 6: headline with floats, 7: pre-net, 8: tunnel and belly, 9: codend.

The body of the trawl is funnel-like, wide at its "mouth" and narrowing towards the cod end, and usually is fitted with wings of netting at the both sides of the mouth. It is long enough to assure adequate flow of water and prevent fish from escaping the net, after having been caught. It is made of diamond-meshed netting, the size of the meshes decreasing from the front of the net towards the codend. Into the body, fish and turtle escape devices can be fitted. These can be simple structures like "square mesh panels", which are easier for smaller fish to pass through, or more complicated devices, such as bycatch grills.

Cod End

The cod end is the trailing end of the net where fish are finally "caught". The size of mesh in the cod end is a determinant of the size of fish which the net catches. Consequently, regulation of mesh size is a common way of managing mortality of juvenile fishes in trawl nets.

How Trawls Work?

The idea that fish are passively "scooped up" is commonly held, and has been since trawling was first developed, but has been revealed to be erroneous. Since the development of scuba diving equipment and cheap video cameras it has been possible to directly observe the processes that occur when a trawl is towed along the seabed. The trawl doors disturb the sea bed, create a cloud of muddy water which hides the oncoming trawl net and generates a noise which attracts fish.

The fish begin to swim in front of the net mouth. As the trawl continues along the seabed, fish begin to tire and slip backwards into the net. Finally, the fish become exhausted and drop back, into the "cod end" and are caught. The speed that the trawl is towed at depends on the swimming speed of the species which is being targeted and the exact gear that is being used, but for most demersal species, a speed of around 4 knots (7 km/h) is appropriate.

Bobbins (lower left) and floats (lower right - upper right) on a trawl net.

Environmental Damage

Trawling gear produces acute impacts on biota and the physical substratum of the seafloor by disrupting the sediment column structure, overturning boulders, resuspending sediments and imprinting deep scars on muddy bottoms. Also, the repetitive passage of trawling gear over the same areas creates long-lasting, cumulative impacts that modify the cohesiveness and texture of sediments. It can be asserted nowadays that due to its recurrence, mobility and wide geographical extent, industrial trawling has become a major force driving seafloor change and affecting not only its physical integrity on short spatial scales but also imprinting measurable modifications to the geomorphology of entire continental margins.

Bottom fishing has operated for over a century on heavily fished grounds such as the North Sea and Grand Banks. While overfishing has long been recognised as causing major ecological changes to the fish community on the Grand Banks, concern has been raised more recently about the damage which benthic trawling inflicts upon seabed communities. A species of particular concern is the slow growing, deep water coral *Lophelia pertusa*. This species is home to a diverse community of deep sea organisms, but is easily damaged by fishing gear. On 17 November 2004, the United Nations General Assembly urged nations to consider temporary bans on high seas bottom trawling. A global analysis of the impacts of bottom trawling found that the impact on seabed biota was strongly dependent on the type of gear used, with otter trawls estimated as having the smallest impact and removing 6% of biota per pass while hydraulic dredges had the largest impact and removed 41% of biota per pass.

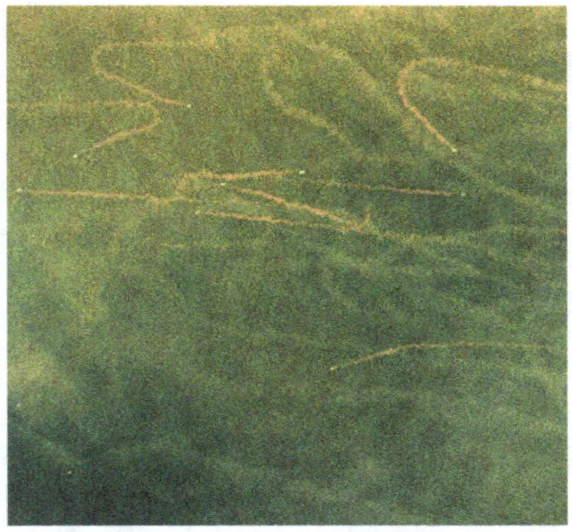

Satellite image of trawler mud trails off the Louisiana coast.

Resuspension

Bottom trawling stirs up the sediment at the bottom of the sea. The suspended solid plumes can drift with the current for tens of kilometres from the source of the trawling, increasing sedimentation rates in deep environments Bottom trawling-induced resuspended sediment mass on the world's continental shelves have been estimated to ~22 Gigatonnes per year, approximately the same as the sediment mass supplied to the continental shelves through the world's rivers. These plumes introduce a turbidity which decreases light levels at the bottom and can affect kelp reproduction.

Ocean sediments are the sink for many persistent organic pollutants, usually lipophilic pollutants like DDT, PCB and PAH. Bottom trawling mixes these pollutants into the plankton ecology where they can move back up the food chain and into our food supply.

Phosphorus is often found in high concentration in soft shallow sediments. Resuspending nutrient solids like these can introduce oxygen demand into the water column, and result in oxygen deficient dead zones.

Even in areas where the bottom sediments are ancient, bottom trawling, by reintroducing the sediment into the water column, can create harmful algae blooms. More suspended solids are introduced into the oceans from bottom trawling than any other man-made source.

Deep Sea Damage

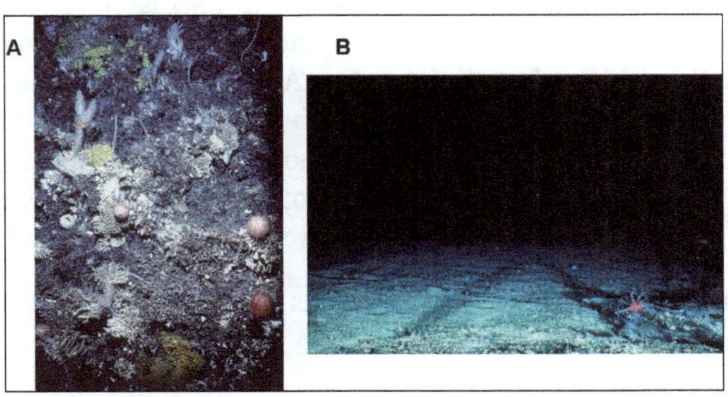

(A): The coral community and seabed on an untrawled seamount. (B): The exposed bedrock of a trawled seamount. Both are 1,000–2,000 meters (1094–2188 yards) below the surface.

The UN Secretary General reported in 2006 that 95 percent of damage to seamount ecosystems worldwide is caused by deep sea bottom trawling. A study published in *Current Biology* suggests a cutoff of 600 meters is a point which ecological damage increases significantly.

Current Restrictions

Today, some countries regulate bottom trawling within their jurisdictions:

- The United States Regional Fishery Management Councils limit bottom trawling in specific closed areas to protect specific species or habitat. For instance, on the U.S. West Coast a large Rockfish Conservation Area was created in 2002 prohibiting trawling in most areas of the coast between 75 and 150 fathoms to protect overfished rockfish species. In 2018, these closures were revised to allow trawling in some previously closed areas while closing new areas of sensitive habitat to bottom trawling.
- The Council of the European Union in 2004 applied "a precautionary approach" and closed the sensitive Darwin Mounds off Scotland to bottom trawling.
- In 2005, the FAO's General Fisheries Commission for the Mediterranean (GFCM) banned bottom trawling below 1000 metres and, in January 2006, completely closed ecologically sensitive areas off Italy, Cyprus, and Egypt to all bottom trawling.
- Norway first recognized in 1999 that trawling had caused significant damage to its cold-water lophelia corals. Norway has since established a program to determine the location of cold-water corals within its EEZ so as to quickly close those areas to bottom trawling.
- Canada has acted to protect vulnerable coral reef ecosystems from bottom trawling off Nova Scotia. The Northeast Channel was protected by a fisheries closure in 2002, and the Gully area was protected by its designation as a Marine Protected Area (MPA) in 2004.
- Australia in 1999 established the Tasmanian Seamounts Marine Reserve to prohibit bottom trawling in the south Tasman Sea. Australia also prohibits bottom trawling in The Great Australian Bight Marine Park near Ceduna off South Australia. In 2004, Australia established the world's largest marine protected area in the Great Barrier Reef Marine Park where fishing and other extractive activities are prohibited.
- New Zealand in 2001 closed 19 seamounts within its EEZ to bottom trawling, including in the Chatham Rise, sub-Antarctic waters, and off the east and west coasts of the North Island. New Zealand Fisheries Minister Jim Anderton announced on 14 February 2006 that a draft agreement had been reached with fishing companies to ban bottom trawling in 30 percent of New Zealand's exclusive economic zone, an area of about 1.2 million km^2 reaching from sub-Antarctic waters to sub-tropical ones. But only a small fraction of the area proposed for protection will cover areas actually vulnerable to bottom trawling.
- Palau has banned all bottom trawling within its jurisdiction and by any Palauan or Palauan corporation anywhere in the world.
- The President of Kiribati, Anote Tong, announced in early 2006 the formation of the world's first deep sea marine reserve area. This measure— the Phoenix Islands Protected

Area— creates the world's third largest marine protected area and may protect deep sea corals, fish, and seamounts from bottom trawling. However, the actual boundaries of this reserve and what harvest limitations may occur therein have not been detailed. Moreover, Kiribati currently has only 1 patrol boat to monitor this proposed region.

- Hong Kong passed legislation banning trawling on 18 May 2011 in an effort to restore the territory's devastated fish stocks and marine ecosystem. The ban came into effect on 31 December 2012. The government paid HK$1.72 billion to affected trawlers in a buyout scheme. Persons who contravene the ban can be fined or imprisoned under the Fisheries Protection Ordinance (Cap 171).

Lack of Regulation

Beyond national jurisdictions, most bottom trawling is unregulated either because there is no Regional Fisheries Management Organization (RFMO) with competence to regulate, or else what RFMOs that do exist have not actually regulated. The major exception to this is in the Antarctic region, where the Convention for the Conservation of Antarctic Marine Living Resources regime has instituted extensive bottom trawling restrictions. The North East Atlantic Fisheries Commission (NEAFC) also recently closed four seamounts and part of the mid-Atlantic Ridge from all fishing, including bottom trawling, for three years. This still leaves most of international waters completely without bottom trawl regulation.

As of May 2007 the area managed under the South Pacific Regional Fisheries Management Organisation (SPRFMO) has gained a new level of protection. All countries fishing in the region (accounting for about 25 percent of the global ocean) agreed to exclude bottom trawling on high seas areas where vulnerable ecosystems are likely or known to occur until a specific impact assessment is undertaken and precautionary measures have been implemented. Also observers will be required on all high seas bottom trawlers to ensure enforcement of the regulations.

ELECTRIC PULSE FISHING

Electric pulse fishing is a fishing technique sometimes used in trawl fisheries which produces a limited electric field above the seabed to catch fish. The pulse trawl gear consists of a number of electrodes, attached to the gear in the tow direction, that emit short electric pulses. The electrodes replace the tickler chains that are used in traditional beam trawl fishery. The pulse trawl fishery is mainly aimed at flatfish like sole, with by-catch plaice. In addition, the pulse trawl gear is applied in shrimp fisheries on a limited scale. Technically, the use of electricity to catch fish is prohibited in European waters. However, the European Union is able to provide exemptions to this rule since 2007. These exemptions are now mainly used by Dutch and British trawlers in the North Sea.

Electric pulse fishing should not be confused with pulse fishing, which is a fisheries management technique for preventing fish stocks from being overfished by periodically permitting a cycle of fishing followed by a fallow period which allows stocks to reconstitute.

Electric pulse fishing is applied on a limited scale for catching flatfish. In this branch of demersal fishery different techniques can be used for disturbing fish on the seabed and sending them up into the belly of the net. In conventional bottom trawling the trawl doors and foot rope, which consists of steel chains mounted to the bottom of the net opening, disturb the seabed, causing the demersal fish to swim along in the mouth of the net until they succumb and fall into cod end. Electric pulse fishing gear does not involve these chains, but features cables with electrodes consisting of isolated and conductive elements mounted in the dragging direction. A short-lasting electric field is generated to disturb demersal fish without killing or paralysing them, but rather by causing a involuntary muscular contraction.

Electric pulse fishing consumes less energy compared to conventional trawling, due to the lighter trawl, lower fishing speed and lower frictive resistance from the seabed. It therefore requires significantly less fuel, which reduces carbon dioxide emissions and makes it an economically interesting alternative. As there is no "ploughing effect", this method of fishing has a less disastrous impact on the seabed ecosystem . It also involves fewer discards for all species categories that are discarded.

LONGLINE FISHING

Longline fishing, or longlining, is a commercial fishing technique. It uses a long line, called the main line, with baited hooks attached at intervals by means of branch lines called *snoods* (or *gangions*). A snood is a short length of line, attached to the main line using a clip or swivel, with the hook at the other end. Longlines are classified mainly by where they are placed in the water column. This can be at the surface or at the bottom. Lines can also be set by means of an anchor, or left to drift. Hundreds or even thousands of baited hooks can hang from a single line. Longliners – fishing vessels rigged for longlining – commonly target swordfish, tuna, halibut, sablefish and many other species.

In some unstable fisheries, such as the Patagonian toothfish, fishermen may be limited to as few as 25 hooks per line. In contrast, commercial longliners in certain robust fisheries of the Bering Sea and North Pacific generally run over 2,500 hand-baited hooks on a single series of connected lines many miles in length.

Longlines can be set to hang near the surface (pelagic longline) to catch fish such as tuna and swordfish or along the sea floor (demersal longline) for groundfish such as halibut or cod. Longliners fishing for sablefish, also referred to as black cod, occasionally set gear on the sea floor at depths exceeding 1,100 metres (3,600 ft) using relatively simple equipment. Longlines with traps attached rather than hooks can be used for crab fishing in deep waters.

Longline fishing is prone to the incidental catching and killing of dolphins, seabirds, sea turtles, and sharks, but can be considerably more ecologically sustainable than some other commercially significant harvesting methods.

Incidental Catch

Longline fishing is controversial in some areas because of bycatch, fish caught while seeking another species or immature juveniles of the target species. This can cause many issues, such as the killing of many other marine animals while seeking certain commercial fish. Seabirds can be particularly vulnerable during the setting of the line.

Seabirds with longline fishing vessel.

Methods to mitigate incidental mortality have succeeded in some fisheries. Mitigation techniques include the use of weights to ensure the lines sink quickly, the deployment of streamer lines to scare away birds, setting lines only at night in low light (to avoid attracting birds), limiting fishing seasons to the southern winter (when most seabirds are not feeding young), and not discharging offal while setting lines.

The Hawaii-based longline fishery for swordfish was closed in 2000 over concerns of excessive sea turtle by-catch, particularly loggerhead sea turtles and leatherback turtles. Changes to the management rules allowed the fishery to reopen in 2004. Gear modification, particularly a change to large circle-hooks and mackerel-type baits, eliminated much of the sea turtle by-catch associated with the fishing technique. It has been claimed that one consequence of the closure was that 70 Hawaii-based vessels were replaced by 1,500-1,700 longline vessels from various Asian nations, but this is not based on any reliable data. Due to poor and often non-existent catch documentation by these vessels, the number of sea turtles and albatross caught by these vessels between 2000 and 2004 will never be known. Hawaii longline fishing for swordfish closed again on 17 March 2006, when the by-catch limit of 17 loggerhead turtles was reached. In 2010 the by-catch limit for loggerhead turtles was raised, but was restored to the former limit as a result of litigation. The Hawaii-based longline fisheries for tuna and swordfish are managed under sets of slightly different rules. The tuna fishery is one of the best managed fisheries in the world, according to the UN Code of Responsible Fishing, but has been criticized by others as being responsible for continuing by-catch of false killer whales, seabirds, and other nontargeted wildlife, as well as placing pressure on depleted bigeye tuna stocks.

Commercial longline fishing is also one of the main threats to albatrosses. Of the 22 albatross species recognized in the IUCN Red List, six are threatened, and nine are vulnerable. The IUCN lists three species as critically endangered: the Amsterdam albatross, the Tristan albatross and the waved albatross. The remaining four are near threatened. Albatrosses and other seabirds which readily feed on offal are attracted to the set bait, become hooked on the lines and drown. An estimated 8,000 albatross per year are killed in this way.

Black-browed albatross hooked on a long-line.

Safety

In the US, a study found that the risk for non-fatal injuries was 35 per 1,000 full-time equivalent employees, about three times higher than average U.S. worker. (This is compared to 43 per 1,000 in their trawler fleet).

GILLNETTING

Oil painting of gillnetting.

Gillnetting is a common fishing method used by commercial and artisanal fishers of all the oceans and in some freshwater and estuary areas. Gill nets are composed of vertical panels of netting that hang from a line with regularly spaced floaters that hold the line on the surface of the water. The floats are sometimes called "corks" and the line with corks is generally referred to as a "cork line." The line along the bottom of the panels is generally weighted. Traditionally this line has been weighted with lead and may be referred to as "lead line." A gill net is normally set in a straight line. Gill nets can be characterized by mesh size, as well as colour and type of filament from which they are made. Fish may be caught by gill nets in three ways:

- Wedged – Held by the mesh around the body.
- Gilled – Held by mesh slipping behind the opercula.
- Tangled – Held by teeth, spines, maxillaries, or other protrusions without the body penetrating the mesh.

Most often fish are gilled. A fish swims into a net and passes only part way through the mesh. When it struggles to free itself, the twine slips behind the gill cover and prevents escape.

Contemporary Canadian commercial salmon bowpicker on trailer. Gillnet is evident on the metal drum in the bow of the boat.

Gillnets are so effective that their use is closely monitored and regulated by fisheries management and enforcement agencies. Mesh size, twine strength, as well as net length and depth are all closely regulated to reduce bycatch of non-target species. Gillnets have a high degree of size selectivity. Most salmon fisheries in particular have an extremely low incidence of catching non-target species.

A fishing vessel rigged to fish by gillnetting is a gillnetter. A gillnetter which deploys its gillnet from the bow is a bowpicker, while one which deploys its gillnet from the stern is a sternpicker.

Fish and Wildlife Service worker on boat checking gillnet full of fish.

Gillnets existed in ancient times, as archaeological evidence from the Middle East demonstrates. In North America, Native American fishermen used cedar canoes and natural fibre nets, e.g., made with nettles or the inner bark of cedar. They would attach stones to the bottom of the nets as weights, and pieces of wood to the top, to use as floats. This allowed the net to suspend straight up and down in the water. Each net would be suspended either from shore or between two boats. Native fishers in the Pacific Northwest, Canada, and Alaska still commonly use gillnets in their fisheries for salmon and steelhead.

Both drift gillnets and setnets have long been used by cultures around the world. There is evidence of fisheries exploitation, including gillnetting, going far back in Japanese history, with many specific details available from the Edo period (1603–1868). Fisheries in the Shetland Islands, which were settled by Norsemen during the Viking age, share cultural and technological similarities with Norwegian fisheries, including gillnet fisheries for herring. Many of the Norwegian immigrant fishermen who came to fish in the great Columbia River salmon fishery during the second half of the 19th century did so because they had experience in the gillnet fishery for cod in the waters surrounding the Lofoten Islands of northern Norway. Gillnets were used as part of the seasonal round by Swedish fishermen as well. Welsh and English fishermen gillnetted for Atlantic salmon in the rivers of Wales and England in coracles, using hand-made nets, for at least several centuries. These are but a few of the examples of historic gillnet fisheries around the world.

Gillnetting was an early fishing technology in colonial America, used for example, in fisheries for Atlantic salmon and shad. Immigrant fishermen from northern Europe and the Mediterranean brought a number of different adaptations of the technology from their respective homelands with them to the rapidly expanding salmon fisheries of the Columbia River from the 1860s onward. The boats used by these fisherman were typically around 25 feet (8 m) long and powered by oars. Many of these boats also had small sails and were called "row-sail" boats. At the beginning of the 1900s, steam powered ships would haul these smaller boats to their fishing grounds and retrieve them at the end of each day. However, at that time gas powered boats were beginning to make their appearance, and by the 1930s, the row-sail boat had virtually disappeared, except in Bristol Bay, Alaska, where motors were prohibited in the gillnet fishery by territorial law until 1951.

A fisherman repairs a gillnet.

In 1931, the first powered drum was created by Laurie Jarelainen. The drum is a circular device that is set to the side of the boat and draws in the nets. The powered drum allowed the nets to be drawn in much faster and along with the faster gas powered boats, fisherman were able to fish in areas they had previously been unable to go into, thereby revolutionizing the fishing industry.

During World War II, navigation and communication devices, as well as many other forms of maritime equipment (ex. depth-sounding and radar) were improved and made more compact. These devices became much more accessible to the average fisherman, thus making their range and mobility increasingly larger. It also served to make the industry much more competitive, as the fisherman were forced to invest more in boats and equipment to stay current with developing technology.

The introduction of fine synthetic fibres such as nylon in the construction of fishing gear during the 1960s marked an expansion in the commercial use of gillnets. The new materials were cheaper and easier to handle, lasted longer and required less maintenance than natural fibres. In addition, multifilament nylon, monofilament or multimonofilament fibres become almost invisible in water, so nets made with synthetic twines generally caught greater numbers of fish than natural fibre nets used in comparable situations.

Nylon is highly resistant to abrasion and degradation, hence the netting has the potential to last for many years if it is not recovered. This ghost fishing is of environmental concern. Attaching the gillnet floats with biodegradable material can reduce the problem. However it is difficult to generalize about the longevity of ghost-fishing gillnets due to the varying environments in which they are used. Some researchers have found gill-nets still catching fish and crustaceans over a year after loss, while others have found lost nets destroyed by wave action within one month or overgrown with seaweeds, increasing their visibility and reducing their catching potential to such an extent that they became a microhabitat used by small fish.

This type of net was heavily used by many Japanese, South Korean, and Taiwanese fishing fleets on the high seas in the 1980s to target tunas. Although highly selective with respect to size class of animals captured, gillnets are associated with high numbers of incidental captures of cetaceans (whales and dolphins). In the Sri Lankan gillnet fishery, one dolphin is caught for every 1.7–4.0 tonnes of tuna landed. This compares poorly with the rate of one dolphin per 70 tonnes of tuna landed in the eastern Pacific purse seine tuna fishery.

Gillnets are sometimes used by fisheries scientists to monitor fish populations.

Legal Status

United Nations General Assembly Resolution 46/215 called for the cessation of all "large-scale pelagic drift-net fishing" in international waters by the end of 1992. The laws of individual countries vary with regard to fishing in waters under their jurisdiction.

Possession of gillnets is illegal in some U.S. states and heavily regulated in others.

Oregon voters had the chance to decide on whether gillnetting will continue in the Columbia River in November 2012 by voting on Measure 81. The measure was defeated with 65% of Oregon voters voting against the measure and allowing commercial gillnet fishing to continue on the Columbia River.

The Columbia River Basin is currently under a management agreement that spans from 2008 to December 31, 2017. This management agreement looks to gather information on fish harvesting through means including gillnets. The parties involved will convene again to decide on further action after the current agreement ends.

The gill-netting season in Minnesota can vary from county to county and the net types used are regulated on a lake by lake basis by the Minnesota Department of Natural Resources.

Virginia's gill-netting season is regulated by the Virginia Marine Resources Commission. During different months of the year, certain rivers have restricted mesh sizes, which vary by location. There have been proposed regulations to shut down drift gill net fisheries whose by-catch numbers (which include dolphins, sea turtles and other marine life) were too high. In 2014, California lawmakers pushed for the banning of gillnet fishing through letters to federal fishing companies. The progress for these regulations have been paused in California mid 2017.

According to the High Seas Fishing Compliance Act from 1996, a permit is require for all commercial fishing vessels that are registered in the United States and under this act, vessels must have a record of all their fishing efforts on the high seas. As of November 2017, there has been a bill introduced to improve the management of driftnets, with gillnets being under the umbrella of this fishing tool. The bill's focus is to ban the use of large-scale nets while supporting the use of alternative methods of fishing to decrease the maximum amount of bycatch. There is also a compensation plan proposed in the bill for fishery participants who stop using large-scale nets.

Selectivity

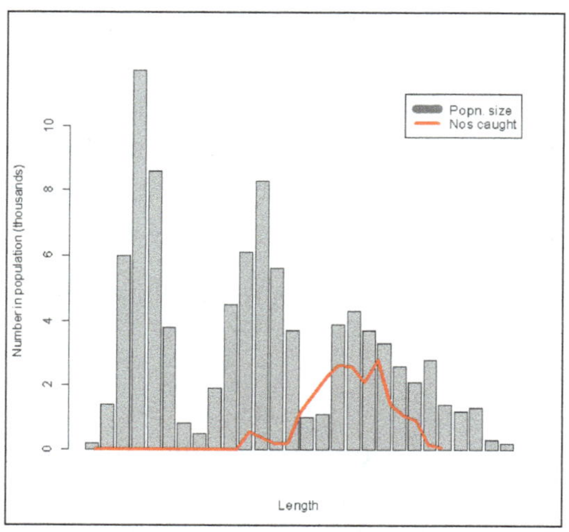

Selectivity properties of a gillnet on a hypothetical population.

Gillnets are a series of panels of meshes with a weighted "foot rope" along the bottom, and a *headline*, to which floats are attached. By altering the ratio of floats to weights, buoyancy changes, and the net can therefore be set to fish at any depth in the water column. In commercial fisheries, the

meshes of a gillnet are uniform in size and shape. Fish smaller than the mesh of the net pass through unhindered, while those too large to push their heads through the meshes as far as their gills are not retained. This gives gillnets the ability to target a specific size of fish, unlike other net gears such as trawls, in which smaller fish pass through the meshes and all larger fish are captured in the net.

Salmon

Commercial gillnet fisheries are still an important method of harvesting salmon in Alaska, British Columbia, Washington, and Oregon. In the lower Columbia River, non-Indian commercial salmon fisheries for spring Chinook have developed methods of selectively harvesting adipose fin clipped hatchery salmon using small mesh gillnets known as tangle nets or tooth nets. Non-adipose fin clipped fish (primarily natural origin salmon) must be released. Fishery management agencies estimate a relatively low release mortality rate on salmon and steelhead released from these small mesh gillnets.

Problems that can arise from selective harvesting are smaller reproducing adult fish, as well as the unexpected mortality of the fish which sustain injuries from the gillnet but are not retained in the fishery. Most salmon populations include several age classes, allowing for fish of different ages, and sizes, to reproduce with each other. A recent 2009 study looked at 59 years of catch and escapement data of Bristol Bay sockeye salmon to determine age and size at maturity trends attributable to the selectivity of commercial gillnet harvests. The study found that the larger females (>550 mm) of all age classes were most susceptible to harvest. The study suggests that smaller, younger fish were more likely to successfully traverse the gillnet fishery and reproduce than the larger fish. The study also found that the average length of sockeye harvested from 1946–2005 was 8 mm larger than the sockeye who escaped the gillnet fishery to spawn, reducing the fecundity of the average female by 5%, or 104 eggs. If a salmon enters a gillnet, but manages to escape, it can sustain injuries. These injuries can lead to a lower degree of reproductive success. A study aimed at quantifying mortality of Bristol Bay sockeye salmon due to gillnet-related injuries found that 11–29% of sockeye sustained fishery-related injuries attributable to gillnets, and 51% of those fish were expected to not reproduce.

Gillnets are sometimes a controversial gear type especially among sport fishers who argue they are inappropriate especially for salmon fisheries. These arguments are often related to allocation issues between commercial and recreational (sport) fisheries and not conservation issues. Most salmon fisheries, especially those targeting Pacific salmon in North America, are strictly managed to minimize total impacts to specific populations and salmon fishery managers continue to allow the use of gillnets in these fisheries.

In 2012, University of Washington Fisheries Professor Emeritus Stephen Mathews compared Puget Sound bycatch data for the non-treaty gillnet and purse seine keta salmon fisheries. He found that although neither fishery had major bycatch problems with nontarget salmonids, the gillnet fishery has substantially less impact on nontarget Chinook salmon. His fulltext report is available from the Washington State Puget Sound Salmon Commission.

Swordfish

Gillnets are also used out in the deep sea for fisheries whose primary catch is swordfish. California driftnet fisheries have some of the highest rates of by-catch with 12 percent of the catch being the targeted swordfish while up to 68 percent of the catch being by-catch that will be tossed back to sea.

Alternatives

Given the selective properties of gillnet fishing, alternative methods of harvest are currently being studied. Recent WDF&W reports suggest that purse seine is the most productive method with having highest catch per unit effort (CPUE), but has little information on the effectiveness of selectively harvesting hatchery-reared salmon. More conclusive research has been conducted jointly between the Confederated Tribes of the Colville Reservation and Bonneville Power Administration on a 10-year study on selective harvest methods of hatchery origin salmon in the Upper Columbia River by purse seine and tangle net. Their 2009 and 2010 findings show that purse seines have a higher percentage of survivability and higher CPUE than does tangle nets. A Colville Tribe biologist reports that during these two years the tribe harvested 3,163 hatchery Chinook while releasing 2,346 wild Chinook with only 1.4% direct or immediate mortality using purse seines, whereas the tangle net was far less productive but had an approximate 12.5% mortality. Researchers commented that the use of recovery boxes and shortened periods between checking the nets would have likely decreased mortality rates. While there is data that shows success of selective methods of harvest at protecting wild and ESA listed salmon, there still must be social acceptance of new methods of fishing.

There have also been studies done to see differing strategies that could potentially decrease avian by-catch in coastal fisheries. These include three strategies that have a possible reduction in up to 75% of avian by-catch: gear modifications, where visual devices will be placed near the top of the net so birds will be able to see the nets; abundance-based fishery openings, where of birds will determine whether the nets will be set out or not; and time-of-day restrictions, which goes along with abundance- where bird by catch tended to occur at dawn and dusk, where as fish catch occurred mostly at dawn.

For marine mammal by-catch, field experiments have shown that the use of pingers on nets resulted in significantly lower numbers of by-catch than nets without pingers. After this study was completed by Jay Barlow, it was determined that there would be a 12-fold decrease in short-beaked common dolphins caught, a 4-fold decrease in other cetaceans and a 3-fold decrease in pinnipeds for nets containing pingers.

Types of Gillnets

The FAO classifies gillnet gear types as follows:

Set Gillnets

Set gillnets consist of a single netting wall kept vertical by a floatline (upper line/headrope) and a weighted groundline (lower line/footrope). Small floats, usually shaped like eggs or cylinders and made of solid plastic, are evenly distributed along the floatline, while lead weights are evenly distributed along groundline. The lower line can also be made of lead cored rope, which does not need additional weight. The net is set on the bottom, or at a distance above it and held in place with anchors or weights on both ends. By adjusting the design these nets can fish in surface layers, in mid water or at the bottom, targeting pelagic, demersal or benthic species. On small boats gillnets are handled by hand. Larger boats use hydraulic net haulers or net drums. Set gillnets are widely used all over the world, and are employed both in inland and sea waters. They are popular with artisanal fisheries because no specialized gear is needed, and it is low cost based on the relationship of fuel/fish.

Encircling Gillnets

A gill netter.

Encircling gillnets are gillnets set vertically in shallow water, with the floatline remaining at the surface so they encircle fish. Small open boats or canoes can be used to set the net around the fish. Once the fish are encircled, the fishers shout and splash the water to panic the fish so they gill or entangle themselves. There is little negative impact on the environment. As soon as the gear is set the scaring takes place and the net is hauled back in. The fish are alive and discards can be returned to the sea. Encircling gillnets are commonly used by groups of small-scale fishers, and does not require other equipment.

Combined Gillnets-trammel Nets

This bottom-set gear has two parts:

- The upper part is a standard gillnet where semi-demersal or pelagic fish can be gilled
- The lower part is a trammel net where bottom fish can entangle.

The combined nets are maintained more or less vertically in the usual way by floats on the floatline and weights on the groundline. They are set on the bottom. After a time depending on the target species, they are hauled on board. Traditional combined nets were hauled by hand, especially on smaller boats. Recent hydraulic driven net haulers are now common. The gilled, entangled and enmeshed fish are removed from the net by hand. Of some concern with this method is ghost fishing by lost nets and bycatch of diving seabirds. Nets combined in this way were first used in the Mediterranean.

Drift Nets

A drift net consists of one or more panels of webbing fastened together. They are left free to drift with the current, usually near the surface or not far below it. Floats on the floatline and weights on the groundline keep them vertical. Drift nets drift with the current while they are connected with the operating vessel, the driftnetter or drifter.

Drift nets are usually used to catch schooling forage fish such as herring and sardines, and also larger pelagic fish such as tuna, salmon and pelagic squid. Net haulers are usually used to set and haul driftnets, with a drifter capstan on the forepart of the vessel. In developing countries most nets are hauled by hand. The mesh size of the gillnets is very effective at selecting or regulating the size of fish caught. The drift net has a low fuel/fish energy consumption compared to other fishing gear. However, the issue of concern with this type of net is the bycatch of species that are not targeted, such as marine mammals, seabirds and to a minor extent turtles. The use of drift nets longer than 2.5 kilometres on the high seas was banned by the United Nations in 1991. Prior to this ban, drift nets were reaching lengths of 60 kilometres. However, there are still serious concerns with ongoing violations.

Gillnets and Entangling Nets

The tangle net, or tooth net, originated in British Columbia, Canada, as a gear specifically developed for selective fisheries. Tangle nets have smaller mesh sizes than standard gillnets. They are designed to catch fish by their nose or jaw, enabling bycatch to be resuscitated and released unharmed. Tangle nets as adapted to the mark-selective fishery for spring Chinook salmon on the lower Columbia River have a standard mesh size of 4-1/4 inches (10.8 cm.). Short net lengths and soak times are used in an effort to land fish in good condition. Tangle nets are typically used in situations where the release of certain (usually wild) fish unharmed is desirable. In a typical situation calling for the use of a tangle net, for instance, all fish retaining their adipose fins (usually wild) must be returned to the water. Tangle nets are used in conjunction with a live recovery box, which acts as a resuscitation chamber for unmarked fish that appear lethargic or stressed before their release into the water.

SEINE FISHING

Seine fishing (or seine-haul fishing) is a method of fishing that employs a fishing net called a seine, that hangs vertically in the water with its bottom edge held down by weights and its top edge buoyed by floats. Seine nets can be deployed from the shore as a beach seine, or from a boat.

Boats deploying seine nets are known as seiners. Two main types of seine net are deployed from seiners: *purse seines* and *Danish seines*.

Seines have been used widely in the past, including by stone age societies. For example, the Māori used large canoes to deploy seine nets which could be over a kilometer long. The nets were woven from green flax, with stone weights and light wood or gourd floats, and could require hundreds of men to haul.

Native Americans on the Columbia River wove seine nets from spruce root fibers or wild grass, again using stones as weights. For floats they used sticks made of cedar which moved in a way which frightened the fish and helped keep them together.

Seine nets are also well documented in ancient cultures in the Mediterranean region. They appear in Egyptian tomb paintings from 3000 BC. In ancient Roman literature, the poet Ovid makes many references to seine nets, including the use of cork floats and lead weights.

Purse Seine

Purse seine.

The net is drawn or "pursed" so it is closed at the bottom as well.

A common type of seine is a purse seine, named such because along the bottom are a number of rings. A line (referred to as a purse-line) passes through all the rings, and when pulled, draws the rings close to one another, preventing the fish from "sounding", or swimming down to escape the net. This operation is similar to a traditional style purse, which has a drawstring. The purse seine is a preferred technique for capturing fish species which school, or aggregate, close to the surface: sardines, mackerel, anchovies, herring, and certain species of tuna (schooling); and salmon soon before they swim up rivers and streams to spawn (aggregation). Boats equipped with purse seines are called purse seiners.

Purse seiner on the U.S. Pacific Coast.

Purse seine fishing can result in smaller amounts of by-catch (unintentionally caught fish), especially when used to catch large species of fish (like herring or mackerel) that shoal tightly together. When used to catch fish that shoal together with other species, or when used in parallel with fish aggregating devices, the percentage of by-catch greatly increases.

A school of about 400 tons of jack mackerel encircled by a Chilean purse seiner.

Use of purse seines is regulated by many countries. In Sri Lanka, using this type of net within 7 kilometers of the shore is illegal. However, they can be used in the deep sea, after obtaining permission from authorities. Purse seine fishing can have negative impacts on fish stocks because it can involve the bycatch of non-target species and it can put too much pressure on fish stocks.

Power block

The power block is a mechanized pulley used on some seiners to haul in the nets. According to the UN Food and Agriculture Organization, no single invention has contributed more to the success of purse seine net hauling than the power block.

Closeup of the power block.

The Puretic power block line was introduced in the 1950s and was the key factor in the mechanization of purse seining. The combination of these blocks with advances in fluid hydraulics and the new large synthetic nets changed the character of purse seine fishing. The original Puretic power block was driven by an endless rope from the warping head of a winch. Nowadays, power blocks are usually driven by hydraulic pumps powered by the main or auxiliary engine. Their rpm, pull and direction can be controlled remotely.

A minimum of three people are required for power block seining; the skipper, skiff operator, and corkline stacker. In many operations a fourth person stacks the leadline, and often a fifth person stacks the web.

Drum

Drum net seiner in Johnstone Strait, British Columbia.

In certain parts of the western United States as well as Canada, specifically on the coast of British Columbia, drum seining is a method of seine fishing which was adopted in the late 1950s and is now used exclusively in that region.

The drum seine uses a horizontally mounted drum to haul and store the net instead of a power block. The net is pulled in over a roller, which spans the stern, and then passes through a spooling gear with upright rollers. The spooling gear is moved from side to side across the stern which allows the net to be guided and wound tightly on the drum.

There are several advantages to the drum seine over the power block. The net can be hauled very quickly - at more than twice the speed of using a power block, the net does not require overhead handling, and the process is therefore safer. The most important advantage is that the drum system can be operated with fewer deckhands. However, it is illegal to use a seine drum in the state of Alaska.

Danish Seine

A Danish seine, also occasionally called an anchor seine, consists of a conical net with two long wings with a bag where the fish collect. Drag lines extend from the wings, and are long so they can surround an area.

A Danish seine is similar to a small trawl net, but the wire warps are much longer and there are no otter boards. The seine boat drags the warps and the net in a circle around the fish. The motion of the warps herds the fish into the central net.

Danish seiner vessels are usually larger than purse seiners, though they are often accompanied by a smaller vessel. The drag lines are often stored on drums or coiled onto the deck by a coiling machine. A brightly coloured buoy, anchored as a "marker", serves as a fixed point when hauling the seine. A power block, usually mounted on a boom or a slewing deck crane, hauls the seine net.

Danish seining works best on demersal fish which are either scattered on or close to the bottom of the sea, or are aggregated (schooling). They are used when there are flat but rough seabeds which are not trawlable. It is especially useful in northern regions, but not much in tropical to sub-tropical areas.

The net is deployed, with one end attached to an anchored dan (marker) buoy, by the main vessel, the seiner, or by a smaller auxiliary boat. A drag line is paid out, followed by a net wing. As the seiner sweeps in a big circle returning to the buoy, the deployment continues with the seine bag and the remaining wing, finishing with the remaining drag line. In this way a large area can be surrounded. Next the drag lines are hauled in using rope-coiling machines until the catch bag can be secured.

The seine netting method developed in Denmark. Scottish seining ("fly dragging") was a later modification. The original procedure is much the same as fly dragging except for the use of an anchored marker buoy when hauling, and closing the net and warps and net by winch.

ELECTROFISHING

Electrofishing uses direct current electricity flowing between a submerged cathode and anode. This affects the movement of the fish so that they swim toward the anode, where they can be caught.

Electrofishing is a common scientific survey method used to sample fish populations to determine abundance, density, and species composition. When performed correctly, electrofishing results in no permanent harm to fish, which return to their natural state in as little as two minutes after being caught.

Method

Electrofishing relies on two electrodes which deliver direct current at high-voltage from the anode to the cathode through the water. When a fish encounters a large enough potential gradient on this path, it becomes affected by the electricity. Usually pulsed direct current (DC) is applied, which causes galvanotaxis in the fish. Galvanotaxis is uncontrolled muscular convulsion that results in the fish swimming towards the anode. At least two people are required for an effective electrofishing crew: one to operate the anode, and the other to catch the stunned fish with a dip net.

Types

There are three types of electrofishers: backpack models, tote barge models, and boat mounted models, sometimes called a stunboat.

Backpack electrofisher generators are either battery or gas powered. They employ a transformer to pulse the current before it is delivered into the water. The anode is located at the end of a long, two meter pole and is usually in the form of a ring. The cathode is a long, three meter braided steel cable that trails behind the operator. The electrofisher is operated by a deadman's switch on the anode pole. There are a number of safety features built into newer backpack models, such as audible speakers that sound when the unit is operating, tilt-switches that incapacitate the electrofisher if the backpack is tilted more than 45 degrees, and quick-release straps to enable the user to quickly remove the electrofisher in the event of some emergency.

Electrofishing by boat.

When boat electrofishing, the boat itself is the cathode, and the anode(s) are generally mounted off the bow. The stunned fish swim toward the anode, where they are caught alive using a dip net. Towed barge electrofishers operate similarly to backpack electrofishers, with the exception that the generator is located on a floating barge instead of on a backpack. Often the barge can be left stationary on the shore and longer cathodes and anodes allow the crew to sample large areas. Barge electrofishers often employ gas-powered generators since a user does not have to carry the extra weight on his or her back.

A method of electrofishing using illegally diverted electricity from overhead electric lines is prevalent in parts of the Indian State of Kerala. This can result in accidental deaths due to electrocution.

Effectiveness

The effectiveness of electrofishing is influenced by a variety of biological, technical, logistical, and environmental factors. The catch is often selectively biased as to fish size and species composition. When using pulsed DC for fishing, the pulse rate and the intensity of the electric field strongly influence the size and nature of the catch. The conductivity of the water influences the shape and extent of the electric field and thus affects the field's ability to induce capture-prone behavior in the fish.

It can also cause injury to the fish. The electricity causes muscle spasms that damage the vertebrae. For unknown reasons, this is more common and severe in longer fish.

The Benefits of Electrofishing

Electrofishing helps biologists track reproductive success and survival rates of fish species. The assessment is performed by sending an electric current into the water in order to safely stun any nearby fish. Stunned fish can then be easily scooped up in a net and placed in a temporary holding tank where they can revive and be observed for data collection. Most often, fish are measured, weighed and marked with PIT or Floy tags, which are used to determine the health and growth of the fish year over year. Then, they are released back into the water completely unharmed.

How does it Work?

Electrofishing requires a crew of properly trained and licensed lake and pond management professionals as well as a specialized boat, which is often custom built. While electrofishing boats can vary in design, they are all outfitted with the same equipment, including a generator and redundant safety features. The generator directs an AC electric current to a specialized box where it is converted to a DC current—which is much safer for the fish—and then to two nodes overhanging the water. The box operator is able to dial in on specific pattern and strength settings based on the conductivity of the particular waterbody.

Benefits of an Electrofishing Assessment

First, this method captures a large and diverse sample of fish in a short period of time, without the need to catch every fish in the waterbody. By using information from this sample, biologists can collect data on species abundance, size structure and condition—which, in turn, can be used to determine which strategies could be implemented to meet the management goals of the fishery.

These strategies may include, but are not limited to fish stocking, harvest restrictions, water quality improvements, fertilization, feeding and fisheries habitat management.

Biologists can use this method to inspect each fish for the presence of sores or parasites, and identify any prevalent problems throughout the fish populations that would otherwise go unnoticed. Electrofishing can also help expose any unwanted or invasive fish species that are causing harm to the beneficial fish populations or impeding the long-term goals of the fishery. In some cases, the removal of nuisance or invasive species is necessary to most effectively meet the management plan of a lake or pond. These unwanted fish can be removed through electrofishing.

An electrofishing assessment should always be considered at the beginning of a new lake, pond or fisheries management plan, and later implemented on a 1- to 5-year basis depending on the goals of the fishery. Lakes and ponds are constantly changing from year to year, and regular assessments will make it possible to consistently implement the best management practices and make necessary adaptations to the plan when needed. So, whether you're looking to manage your body of water for prized catches, casual fishing, or just want to know what kind of fish are present, your lake management specialist can assess the wildlife in your waterbody and help you make the best decisions for the growth and enhancement of your lake or pond.

References

- Trolling, topic: britannica.com, Retrieved 8 January, 2019
- Fao (2007) workshop on standardization of selectivity methods applied to trawling fisheries report no. 820. Isbn 978-92-5-005669-2
- Trolling, capture, fish-welfare-in-commercial-fishing: fishcount.org.uk, Retrieved 13 May, 2019
- Helfman, gene s (2007) fish conservation: a guide to understanding and restoring global aquatic biodiversity and fishery resources island press, pp. 452–453. Isbn 978-1-55963-596-7
- Keys-to-trolling-success: liveabout.com, Retrieved 25 February, 2019
- Clover, charles. 2004. The end of the line: how overfishing is changing the world and what we eat. Ebury press, london. Isbn 0-09-189780-7
- "fishing and farming methods from the seafood watch program at the monterey bay aquarium". Www.seafoodwatch.org. Retrieved 2018-08-23
- Gabriel o., von brandt a., lange k., dahm e. And wendt t. (2005) fish catching methods of the world: seining in fresh and sea water wiley-blackwell, page 431–448. Isbn 978-0-85238-280-6

Fishery Monitoring, Control and Surveillance

CHAPTER 4

There are different kinds of systems used for monitoring and controlling a fishery. Vessel monitoring system, fishery resource monitoring system and catch reporting are a few parts of its monitoring and control system. The applications of these systems have been thoroughly discussed in this chapter.

FISHERIES MONITORING

Monitoring must be considered within the broader context of fisheries management, which in turn is a part of a mosaic of multiple ocean and natural resource uses, including environmental conservation. Monitoring has a role to play in all aspects of fisheries management, including those related to the sustainable management of the resource, the economic performance of the fishery and the distribution of benefits from the exploitation of the resource and use of the environment. An immense challenge faced by management is that of finding cost-effective monitoring methods. Given its multiple roles, resources must be allocated to monitoring against competing but often related needs from other management related activities such as licensing, planning and legislation, policy formulation, marketing, surveillance and enforcement and research. The challenges of fisheries monitoring and the extent to which solutions lie in (i) finding appropriate fisheries management models and plans, (ii) understanding the fishers motivation for compliance, and (iii) technical solutions.

Fisheries are just one of the users of ocean and aquatic system resources, albeit the user considered to be having the greatest impact on the oceans. Oceans and resources management form the broader frame for fisheries management, and, within this, monitoring, control and surveillance are key functions. Monitoring as a function, therefore, must be integrated with other functions and activities which are much more encompassing in their scope. Indeed, monitoring is only relevant in the larger context of fisheries and oceans management.

By integrated fisheries management we refer to the rational exploitation of fisheries resources based on notions of sustainability efficiency and equity.

We take our definitions of monitoring, control and surveillance (MCS) from those established and accepted at the 1981 International MCS Symposium sponsored by the Food and Agriculture Organization of the United Nations:

1. Monitoring involves the collection, measurement and analysis of fishing activity, including: catches, species composition, fishing effort, discards, area of operation, etc.

2. Control involves the specification of terms and conditions under which resources can be harvested, which are normally contained in fisheries legislation and other arrangements (national, sub-regional, regional).

3. Surveillance involves the checking and supervision of fishing activity to ensure that national legislation and terms, conditions of access, and management measures are observed.

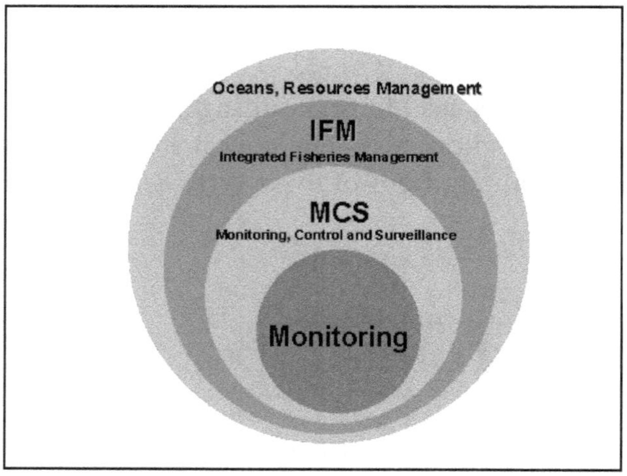

Fisheries management and monitoring within the wider context of oceans management.

Monitoring fisheries operations to assist fisheries management faces formidable technical challenges due to the combined vagaries of a fluctuating and cryptic target resource in its dynamic environment and a mobile exploitation activity. In the majority of fisheries, especially those in tropical developing countries, the large number of species involved, the multiplicity of fishing gears and the widely dispersed landing sites make monitoring, enforcement and compliance measures extremely difficult. Huge costs and already overburdened institutional mechanisms are entailed. High seas and straddling stocks fisheries present extra difficulties and these constraints have paved the way to looking at less costly and more innovative monitoring solutions.

In short, fisheries MCS programmes are required to be effective and efficient within the operational capacity of those charged with their implementation. They must also have legitimacy and, above all, must by relevant and integrated with the fishery being managed.

Fisheries monitoring is undertaken for several purposes. It can help detect extreme events and trends such as the violation of standards, the extent of discards and bycatch. It helps to establish baseline data from which to measure changes. Monitoring also forms the basis for assessing current conditions and helps establish cause-effect relations. Finally, one of the most important purposes of monitoring is to inform in the resolution of fisheries conflicts.

Monitoring data can be many and varied and could include: fish catches, fisheries habitats, number of fishers, census of gear, verification of fishing vessel registration, catch verification and activities that have displaced fisheries activities (e.g. pollution, cages/large fishpens). Monitoring data, the facility with which it may be gathered and its costs, are highly dependent not only on the management regime but also on the nature and scale of the fishery. Thompson demonstrated some of the differences in a table from which figure is extracted.

	LARGE SCALE	SMALL SCALE
Number of fishermen employed	AROUND 500,000	OVER 12,000,000
Annual catch of marine fish for human consumption	AROUND 29 MILLION TONNES	AROUND 24 MILLION TONNES
Capital cost of each job on fishing vessels	$ 30,000-$ 300,000	$ 250-2,500
Fishermen employed for each $ 1 million invested in fishing vessels	5-30	500-4,000
Fish destroyed at sea each year as by-catch in shrimp fisheries	6-16 MILLION TONNES	NONE

Some dimensions of the differences between small and large scale fisheries, at a global scale.

Integrated Fisheries Monitoring

To be effective, fisheries monitoring and enforcement must be integrated within fisheries management and fisheries management integrated within the larger scheme of marine resource and environment management. Put another way: "If you can't measure it, you can't manage it." Fisheries monitoring and enforcement has two essential components: (a) the acquisition and collation of information about fishing activities; and (b) the inducement, on the basis of information obtained, of compliance by fishing vessel operators with desired modes of conduct.

Fisheries monitoring and the collection of data to provide scientific advice for management are inextricably linked in fisheries management and to management in other sectors. Significant attempts have been made through the Code of Conduct for Responsible Fisheries to integrate the various elements of fisheries and aquaculture for sustainable practices. Monitoring and the cost-effective collection of good fisheries data are vital parts of the integration and are mentioned in Article 7.7 of the Code itself and in Article 2.4.4 of the Technical Guidelines for fisheries management under the Code.

Furthermore, if not done effectively, monitoring can even contribute to the failure of fisheries management. For example, 3 out of 7 hypotheses for management failure are related directly or indirectly to monitoring, namely data uncertainty, simple models and complexity.

Fisheries face increasingly complex problems and these cannot be resolved without good monitoring information. For example, the overall catching capacity of commercial fishers has increased tremendously due to improved fishing gear and an increase in the number of boats in commercial fisheries. A large proportion of the catch, that is, about a third of all landings, are discarded. By-catches amount to 27 million annually. In addition, as stocks have become more stressed, small and immature fish are being targeted. To monitor these trends requires data collection that addresses the questions of bycatch and discards explicitly, in addition to the usual collection of retained catch data.

Outside the fisheries sector, most aquatic resource systems are subject to many different uses which place various pressures on them. These may result in lower water quality, decreases in system productivity, and outbreaks of toxic algal blooms which make fish unsafe for human consumption. Monitoring systems must be integrated into such multi-functional frameworks to protect human life and the sustainability of the aquatic resource base.

A further consequence of the multiple uses of aquatic environments is that they are managed by and are of interest to, a host of agencies. The various players have a stake in the effective exploitation, conservation and management of fishery stocks. In every country, a host of government agencies are given marine and fisheries mandates which sometimes overlap. These include agencies whose functions are related to environment and natural resources, agriculture and fisheries, science and technology, minerals and petroleum, economic planning and finance, transportation and communications, national defense, maritime police, local government and foreign affairs. Furthermore, non-government organizations (NGOs) and the private sector are increasingly getting involved in the matter of governance and management of fisheries resources. Integrating even the minimum amount of monitoring information collected across such diverse agencies is an enormous challenge. A first step would be to identify the key data sets to be integrated.

With the increasing pressures on fisheries, fisheries monitoring can entail heavy costs and this at a time when the profitability of fisheries is usually reduced through heavy fishing. Clearly, MCS costs should naturally not exceed national revenue for the fishing industry. Nevertheless, costs of MCS can approach and even exceed these values. In one fishery in the United States, for instance, the national agency spent $80 million on monitoring and surveillance of fishing operations which only collected $51.5 million. In another case, the cost of a modern enforcement programme for the tuna fishery of a small coastal state like Costa Rica was estimated at approximately 50% of the expected revenues.

Furthermore, costs of monitoring have increased sharply as fisheries have become more heavily exploited. In Malaysia, the costs of monitoring, enforcement and compliance have increased from US$3.1 million in 1976-80, to US$16.63 million in 1986-90, and to US$32.2 million in 1991-1995; these correspond to 5%, 16% and 21%, respectively, of total fisheries expenditure.

Considering the links between research data collection and monitoring, we note that much fisheries resources monitoring is carried out through high technology research. For example, oceanographic fieldwork has traditionally meant going to sea in ships. In recent years, it has expanded to include such activities that may require a ship for a short period but then continue independently. The value of programme and equipment integration has been recognised in the United States where, between 1984 and 1997, the University-National Oceanographic Laboratory System undertook a major fleet modernisation and replacement programme. No single agency could have contemplated any part of the programme in isolation.

Monitoring is important to help prevent failure of fisheries management that faces increasingly complex problems. However, given the potentially high costs of monitoring, solutions lie in choosing appropriate fisheries management methods which can be efficiently and effectively monitored, a better understanding of fishers motivation for compliance to the management arrangements and technical solutions which can lower the costs and produce more accurate monitoring information.

Frameworks and Approaches to IFM: Management-oriented Paradigm

Integrated fisheries monitoring (IFM), as with all functions of fisheries management, including research, is dictated by the type of management system in place and must be changed when the system management changes. A good example of how fisheries management changes is when a change in fisheries monitoring occurs in response to fisheries management adopting an individual transferable quota scheme after managing by a total allowable catch. The collection of monitoring data will shift from a focus on overall landings, say at the landing ports, to greater attention to individual landings and trading in quotas.

Just as de la Mare recommended that fisheries science generally should be considered within a management oriented paradigm (MOP), so too we consider that fisheries monitoring must be considered within a MOP. This involves comprehending the whole system of fisheries and their management institutions, not just its parts such as monitoring, the resource and the economics of the fleets. Management objectives and procedures have traditionally been viewed as outside the purview of fisheries science. Smit lamented that fisheries science did not extend to cover the scientific study of management, warning that such a lack of focus on the whole of fisheries would leave fisheries science ineffective in the future as in the past. So too with fisheries monitoring.

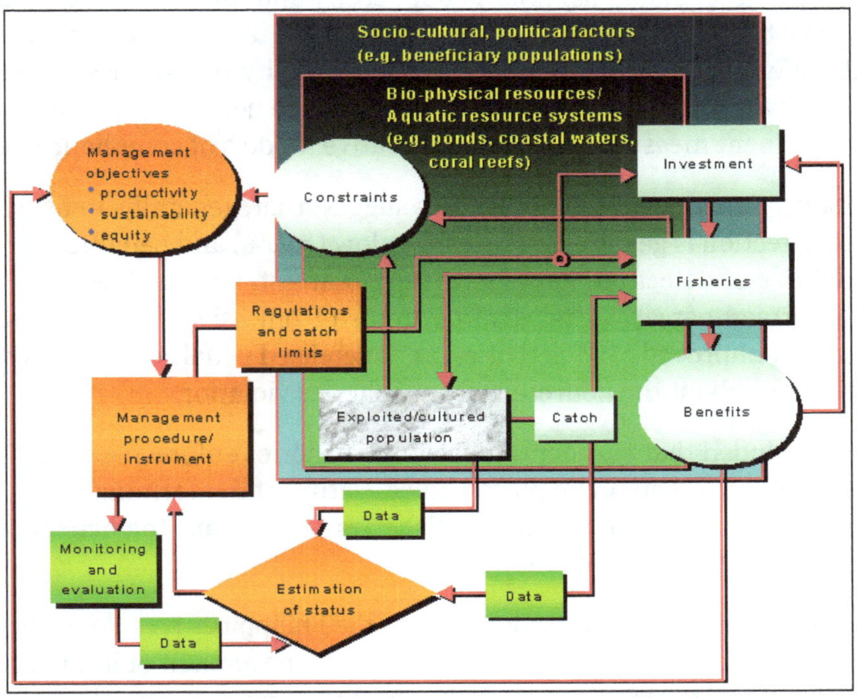

Management oriented paradigm showing the relationship between functions in fisheries management.

The MOP approach for any function consists of: measurable management objectives; a management procedure based on decision rules; assessments based on specific data and methods and a prospective evaluation of the management procedure using performance measures. The fisheries system needs to be studied as a whole, not just as its parts. As with any other system or structure, sustainable management systems have to be designed to get the most benefits at commensurate costs.

When applied to monitoring functions instead of research, other management-related activities need to be considered, including licensing, planning and legislation, policy formulation, marketing, surveillance and enforcement. Monitoring systems, compliance behavior and legitimacy have to be taken into account. Furthermore, monitoring systems need to be codified in law to provide full legitimacy. Good compliance by fishers can make a big difference to the cost effectiveness of monitoring. Conversely, monitoring can be designed to check for compliance. In either case, an understanding of what leads to compliance can be important to management and monitoring.

Compliance Behavior and Legitimacy

Potentially, a management-oriented approach for fisheries activities should positively influence fishers' compliance behavior. Fishers make interesting subjects for the study of compliance. They are subject to numerous regulations and constraints, yet they offer opportunities, often out of sight of others and the law. Fisheries violations are often the result of deliberate choice. Therefore, the behavior of fishers offers good cases with which to test the role of deterrence, legitimacy and other factors that explain compliance, as well as to suggest better and more targeted monitoring and surveillance systems.

Achieving compliance in regulated fisheries is both costly and difficult. Expenditures on enforcement commonly constitute the largest cost element in fisheries management. Low rates of compliance and high enforcement costs can threaten the viability of fisheries regulation and make accurate monitoring very difficult. Such difficulties justify the need to assess the cost-effectiveness of traditional enforcement measures, and test alternative models of compliance behavior.

According to Parsons, Canadian and other studies suggest that, generally, in fisheries, crime pays. The probability of detection is generally perceived as low (<10%) and penalties are low. Hence, the combined probabilities of detection, apprehension and arrest, and the likely penalty to be levied are so low that fishermen perceive a positive net return from fisheries violations. The combined effect of increased and improved MCS activities, and increased penalties should, over time, change fishermen's perceptions about the net return from fisheries violations.

Further, Parsons concluded that another dimension to fisheries enforcement, too often ignored, is education. Greater emphasis should be placed on educating and informing fishermen, processors and the judiciary of the rationale for particular fisheries regulations, the benefits to all of compliance and the serious effects of overfishing.

Studies on fishers operating in Malaysia, Indonesia and the Philippines postulated that four factors are likely to affect compliance decisions by the individual fisher; the amount of illegal gain or benefit; the expected penalty; moral obligation and social influence. In any typical population of fishers, some 5% to 10% of fishers tend to violate chronically and flagrantly and enforcement using coercive methods is the only control mechanism in this case. Non-compliance by the vast majority of fishers is on a small-scale and is often inadvertent or opportunistic and can be controlled by non-coercive methods.

Kuperan and Sutinen found that the basic deterrence model for compliance behavior is a weak model. They suggested that stronger alternatives are 'moral legitimacy' and 'social influence', or institutional or community-based arrangements that ensure compliance. The 5% to 10% 'hard-core' violators tended to be more alienated from the fishers' community (e.g. migrants, new entrants, non-fishers who take it up part-time).

Community based Approaches

In many developing countries in recent years, we have witnessed a trend towards community based approaches in traditional or artisanal fisheries. These approaches appear to overcome some of the shortcomings of the traditional deterrence model by focusing on social influence as well as education. The idea of monitoring fisheries resources and controlling the catch is well developed within many community based systems where the lack of formal management laws or regulations does not necessarily mean open access regimes are in place. Rather, customary regimes and other local regimes are commonplace.

Community based systems operate on the following precepts: (a) fishing communities are the primary stakeholders; (b) traditional resource users have property and management rights; (c) self-regulation is the norm; and (d) indigenous methods are in place. The traditional schemes are being integrated with national level systems in fisheries co-management, and with adoption of monitoring and surveillance practices as well. One example is the Orion community based coastal resource management scheme in Bataan, Philippines. Monitoring is done through a fisherfolk based patrol, in collaboration with the municipal police. The *bantay dagat* (sea guard) team has a small patrol boat, from which it monitors fishers activities at sea. Revenues from fines are allocated to fund patrolling expenses.

In open inland water systems in Bangladesh, local management groups are organised to manage oxbow lakes fisheries in western Bangladesh. All fishers according to gear type in the group operate as teams and elect a leader. Each year, a committee is formed from these leaders, and is responsible for the operation and unity of the group.

'No Force' Approach

Another approach, called the 'no force' approach, is used by the South Pacific Forum Fisheries Agency. This is a low cost approach, especially with respect to monitoring, and is now being applied in other parts of the world. Licensing of foreign fishing operations is a significant dollar earner in many developing countries, especially small island developing states (SIDS), but net benefits to the coastal state has not kept pace with increased costs to the fished state and increased fees to the distant water fleets. Substantial costs to both sides are involved in policing fishing operations in exclusive economic zones (EEZs). Surveillance and physical enforcement costs can easily outstrip the financial benefits derived from licensing and, faced with this dilemma, many SIDS have sought new techniques to ensure compliance with their national laws by foreign vessels. Significant progress has been made in developing cooperative means of compliance control particularly at the regional level.

Such a 'no force approach' was established in 1986 by the South Pacific Forum as a regional MCS scheme. A 'register' collates the following information on a region-wide basis: (a) foreign fishing vessels operating in the region; (b) their characteristics and catching potential; (c) their historical catches and (d) fishing areas. Access arrangements and management systems are then designed based on catching potential rather than the actual catch, thus reducing pressures on the information-gathering and verification process. Primary reliance is based on self-reporting by the vessels themselves. With respect to compliance control, cooperative efforts are in place, e.g., regional cost-sharing where physical enforcement is still necessary. `No-force' enforcement methods include: 'good standing' and the threat of blacklisting of the complying and erring fleet, respectively, and the 'flag state responsibility' approach.

The register is an invaluable databank that can be used for the purpose of monitoring and economic research on the fishing fleets operating in the region and as a source of information for national fisheries administrators faced with licensing decisions for individual vessels. It also provides the whole basis for surveillance operations under regional defense and surveillance cooperation programmes. The register is now the centerpiece for regional cooperation and enforcement in the South Pacific region. An innovative Vessel Monitoring System (VMS) tracks the movement of fishing vessels.

In the past half-decade, however, the Pacific island countries have changed the development strategy for their tuna fisheries from a rent maximization approach to an approach that entails developing their own "locally-based" tuna industries. How this new development strategy could affect the future of regional cooperation and monitoring should be further explored .

Advocacy Approach

Advocacy approaches to monitoring and surveillance have been demonstrated in several fisheries over the years, notably by activist NGOs such as *Greenpeace*. The catchy messages and information produced by these groups capture the public's imagination and are often powerful in revealing certain fishing practices, in changing the views of the public, and thereby the positions of the industry and government on fisheries management. In August 1996, for instance, *Greenpeace* raised public awareness on the extent of overfishing in North Pacific fisheries, and how "factory trawlers" have been "wasting away" 700,000,000 lb. of fisheries products a year. These advocacy's are usually accompanied by well-planned media campaigns.

Oceans Policy Approach

Several countries are changing their whole marine management regime through higher level oceans policies. These have a strong flow-on effect to fisheries management and monitoring and can assist the integration of fisheries monitoring.

Australia is moving towards a more integrated approach to maritime monitoring and surveillance with the release of its 1998 Oceans Policy. The policy introduces Regional Marine Plans, provides more support for civil society surveillance and awareness, and provides for more fisheries officers, amended fisheries laws and efforts to reduce incursions by foreign fishers.

In April 1995, Canada's Department of Fisheries and Oceans (DFO) merged with the Canadian Coast Guard (CCC) creating one of the largest civilian marine organisations in the world. The new vision of the restructured DFO is "To be a world leader in ocean and aquatic resources management." The 1997 Oceans Act required DFO to lead the development of a National Oceans Management Strategy. In fisheries, this builds on an already strong base of MCS, including extensive at-sea observer programmes and a strong scientific programme as described by Parsons.

A suitable framework for fisheries monitoring is a management -oriented paradigm for the fishery. This approach allows integration of monitoring and other systems such as research into the management arrangements and shows how their design and effectiveness largely depends on the management model adopted. Compliance studies indicate where management regimes fail and how greater use of social and other non-traditional deterrence measures are recommended. Community based approaches, `no-force' regional arrangements and full national MCS systems integrated into a complete oceans or marine policy all assist fisheries management's tasks.

Technical Solutions

The frameworks of regional, national and community-based management described above are all reliant on many technologies for their implementation. Likewise, a change in the technical capability can provide new opportunities for better monitoring and/or more cost effective monitoring of national, regional or local fisheries.

One method for integrated monitoring of the state of fisheries has been the use of global and national stock assessments and status reports. Such reports as the FAO State of World Fisheries and Aquaculture, the global assessment of by-catch and discards, the US annual stock assessments, and the Australian status reports are examples of high level monitoring tools which provide managers, politicians, fishers and the public with a report card on the state of the fisheries.

These reports would not have been produced without the many stock assessments now made possible by greater research efforts and with greater access to computing and electronic communication around the world.

Sometimes a technical fix to a fisheries management problem can change the scale of monitoring challenges drastically. One such case in point is that of the eastern Pacific tuna-dolphin problem. Dolphin bycatches in tuna fisheries have been at international centre stage in the past decade due to keen US interest plus a strong environmental constituency. In the eastern Pacific, tuna schools are closely associated with dolphins. In the 1950s and 1960s, bycatches of dolphins by tuna vessels amounted to several tens of thousands of dolphins per year. Due to subsequent technological improvements by the industry and regulatory action both by the US and the international community, however, dolphin mortality dropped dramatically to 3,600 in 1993.

As many governments seek to bring down the costs of fisheries management, monitoring costs are continually under review. Cost-effective solutions to monitoring complex fisheries are continually being sought.

One field of monitoring which has been quite successful in developing robust, scientifically valid and cost-effective monitoring methods is that of the assessment of the status of coral reefs and their fisheries. These are particularly difficult and complex systems to describe. High-powered and expensive research survey methods have been used as the basis for developing simplified approaches requiring fewer resources and less time to accomplish useable survey results. One of these is the ICLARM Aquanaut survey method, based on several more complex methods. In the Aquanaut method, professional divers undergo two to three days survey training and are certified on the basis of this training to undertake surveys. Data are compatible with the global coral reef database, ReefBase. Before the development of these and other methods, monitoring of coral reef fisheries and habitats was not possible in any cost-effective way, depending as it did on a few well-studied reefs which were the sites for scientific research.

The realisation of the overall impact of trawling on fisheries stocks and habitats has led to more monitoring of its effects in many countries. Any changes in trawling practices will lead to major changes in monitoring efforts related to this relatively unselective method of fishing which is one of the most pervasive. Trawling affects an area equal to half (53%) of the world's continental shelves each year according to the most recent study by Watling and Norse, as cited in a news release by the *Ocean Voice International*. Trawl gear not only catches marketable species but also kills non-target species and can damage the sea floor. The fishing industry and researchers are designing and testing gear and fishing methods with low habitat damage (e.g. fish traps, longlines) and more environmentally friendly and selective trawl gear. Industry-developed by-catch reduction devices will benefit from the cooperation of industry, research and management personnel. Greater emphasis on protected areas management, however, is leading to the establishment of no-trawling zones in some fisheries, to permit recovery of seabeds and to protect biodiversity.

We still lack cost-effective methods for monitoring many small-scale fisheries in both developed and developing countries. Little attention has been paid to developing these methods in recent times, even though the plight of the small-scale fisheries, especially in developing countries, is worsening. Even trawl fisheries data are frequently missing or stored and not analysed in many countries. In Asia, a seven-country project led by ICLARM is now seeking to provide better database storage and analytical tools for the monitoring of trawl fish stocks. Extensive use is being made of computers and the Internet to conduct the project.

CATCH REPORTING

Catch reporting is a part of monitoring control and surveillance of Commercial fishing. Depending on national and local fisheries management practices, catch reports may reveal illegal fishing practices, or simply indicate that a given area is being overfished.

Manual Catch Reporting

The general industry practice is to write out a catch report on paper, and present it to a fisheries management official when they return to port. If information does not seem plausible to the official, the report may be verified by physical inspection of the catch. Alternatively, a suspicious vessel may need to carry an independent observer on future voyages.

Semi-automated Catch Reporting

Some Vessel monitoring systems have features that collect, from keyboard input, the data that constitutes a catch report for the entire voyage. More advanced systems periodically transmit the current catch as electronic mail, so fisheries management centers can determine if a controlled area needs to be closed to further fishing.

While there is no standardization as yet for catch reports, a starting point came from a 1981 Conference of Experts:

- Catch on entry to each controlled area,
- Weekly catch,
- Transshipment,
- Port of landing,
- Catch on exiting a controlled area,
- Days at sea,
- Daily time at sea,
- Seasonal catch limits,
- Per-trip catch limits,
- Limits on catch within certain areas,
- Individual (vessel) transferable quotas,
- Minimum or maximum fish (or shellfish) sizes.

This was extended, in 1993, to include: to include the measurement of:

- Catch,
- Species composition,
- Fishing effort,
- Bycatch (i.e., species unintentionally caught, such as dolphins in tuna fishery),
- Area of operations.

A number of programs require tracking of days at sea (DAS) for a given vessel. They may require tracking the total cumulative catch of a given fishery.

Major Trends

Where the local fishery economy permits, perhaps with international funding, near-real-time catch reporting will become a basic feature of vessel management systems. Software at fisheries management centers will cross-correlate VMS position information, catch reports, and spot inspection reports.

FISHERIES OBSERVER

A fisheries observer is an independent specialist who serves on board commercial fishing vessels, or in fish processing plants and other platforms, and is employed by a fisheries observer program, either directly by a government agency or by a third party contractor. Observers spend anywhere from 1 day to 3 months out at sea before returning to be "debriefed". A debriefing consists of reviewing any unusual occurrences or observations, violations observed, and any safety problems or other hardships they endured during the trip. These data are then integrated into the regional agency's database used to monitor fish quotas.

Observers are usually the only independent data collection source for some types of at-sea information, such as bycatch, catch composition, and gear configuration data. Independent data collection in this context refers to data that is not potentially biased by the fishermen. Creel surveys, trip reports, and other data obtained directly from fishermen can have some dependent bias associated with it. Fisheries-dependent information is critical for the responsible management and conservation of living marine resources, and many worldwide marine resource management regimes utilize fisheries observers for the collection of this data.

The integrity of a fisheries observer program is a function of the conduct, morale, and performance of its employees. Moreover, the stature and stability of a program has direct bearing on the quality of its data products and on the level of confidence that scientists, managers, and policy makers are able to ascribe to the use of this data.

Fisheries Observer Program

Hiring Requirements

A fisheries observer program is responsible for providing the training and support necessary for deploying observers on board fishing vessels in order to collect the fisheries-dependent information essential to achieving the management objectives of the programme. Generally, fisheries observers are required to have an educational background in the biological sciences and a proven ability to handle life at sea. Specific programs have additional or less requirements. Also required is an independent attitude, an ability to get along with people in tight quarters and strong personal integrity. "Hang Loose" and "Stand By" are the general mottos for observers because conditions, boats and circumstances often change at the last minute.

Observer training.

Objectives

Fisheries observer programmes vary according to the management objective. The three main objectives are:

- Scientific: Including stock (targeted and / or bycatch species) assessments and predictions of future fisheries exploitation of stocks. Observer duties include collection of catch composition and environmental / ecological information, and biological sampling of catch.
- Monitoring control and surveillance (MCS): Including assessments of fishing and / or transhipment activities to ensure fisheries management measures are followed. Observer duties include verification of logbooks with fishing and / or transhipment activities and registering compliance with all regulations.
- Fisheries: Including the objectives found in both scientific and MCS observer programmes. Observer duties also include the observer duties found in both scientific and MCS observer programs.

The International Fisheries Observer and Monitoring Conference develops, promotes and enhances effective fishery monitoring programs to ensure sustainable resource management throughout the world's oceans. Their mission is to improve fishery monitoring and observing programs worldwide through sharing practices and development of data collection and analysis. Since many observing entities are contracted, this conference allows dialog between agencies and between those that rely upon the data collected by observers.

VESSEL MONITORING SYSTEM

Vessel Monitoring Systems (VMS) is a general term to describe systems that are used in commercial fishing to allow environmental and fisheries regulatory organizations to track and monitor the activities of fishing vessels. They are a key part of monitoring control and surveillance (MCS) programs at national and international levels. VMS may be used to monitor vessels in the territorial waters of a country or a subdivision of a country, or in the Exclusive Economic Zones (EEZ) that extend 200 nautical miles (370.4 km) from the coasts of many countries. VMS systems are used to improve the management and sustainability of the marine environment, through ensuring proper fishing practices and the prevention of illegal fishing, and thus protect and enhance the livelihoods of fishermen.

The exact functionality of a VMS system and the associated equipment varies with the requirements of the nation of the vessel's registry, and the regional or national water in which the vessel is operating. Within regional and national VMS initiatives there are also sub-divisions which apply different functionality to different vessel categories. Categories may be size or type of vessel or activity. For example:

- Local/regional fish such as scallops in the Northeast U.S., anchovies in Peruvian waters, or rock shrimp in the Gulf of Mexico.
- Highly migratory species (HMS) such as tuna and billfish, or Patagonian toothfish (*Dissostichus eleginoides*) in the Antarctic which can be caught in multiple regions.

VMS describes the specific application of monitoring commercial fishing boats. It is not to be confused with VTS which is describes the specific application of monitoring marine traffic primarily for safety and efficiency in ports and busy waterways. It is also not to be confused with specific communication technologies such as AIS, Iridium, Inmarsat, Argos, GPRS which relate to the communication method on which data is transmitted. Different VMS systems will use different communication technologies depending on the functionality requirements imposed by a national or regional VMS initiative.

The cost of VMS components will vary according to the functionality requirements of the specific system being implemented. In general the higher the functionality the more expensive the equipment and required data link (airtime costs). The cost of a VMS system therefore varies and thus the level of government subsidy (if any) varies according to national and regional requirements. EU and US VMS systems require expensive onboard equipment and large amounts of data to be transmitted over satellite link resulting in high airtime charges, but also provide a very high level of functionality. In other regions where per vessel cost and huge fleet sizes are an issue, communication technologies such as AIS are used which significantly reduce equipment and airtime costs whilst delivering acceptable basic VMS system functionality.

Applications

VMS is intended principally for fisheries management, but the country using it may use the data for other purposes.

Among all the most basic purposes is to monitor the movement of VMS-equipped vessels with respect to restricted fishing areas. A given vessel may have approval to fish in a restricted area, to transit through it without fishing, or it may not be allowed in the area.

Catch Reporting

Not all VMS systems are required to record and transmit catch reports. On the systems with separate PCs, it is reasonably easy to distribute separate software, although the fishermen may or may not be able to install it without assistance and additional equipment. For VMS with dedicated PCs, additional software can be made part of software support with established channels.

In the EU and US there is a distinct trend to making catch reporting part of an overall MCS program, although current VMS systems rarely integrate this capability with position reporting. One system that does to this is Fulcrum Maritime Systems Limited's Vessel Tracking Service which uses the latest data rich satellite application technology to enable electronic catch reports to be sent from the vessel to the VMS directly and automatically sends the report and vessel position data to the appropriate RFO or Regulatory Fishing Body, such as the NEAFC.

Under the European Union scheme, vessels are generally required to report:

- Catch on entry,
- Weekly catch,
- Transshipment,
- Port of landing,

- Catch on exit.

Within the EU an Electronic Reporting System is being implemented as part of the EU VMS system to automate collection of catch data, and exchange of data between EU states. A number of programs require tracking of days at sea (DAS) for a given vessel. They may require tracking the total cumulative catch of a given fishery.

VMS is a planning and analysis tool as well as an aid to operations. Treated as a research database, the cumulative position reports gives an analysis of fishing vessel tracks in search of fish. Coupled with species-specific fishing licenses and catch reports, fisheries managers can estimate the amount of a given fish in an area, the amount taken by fishing vessels, and project overfishing before it happens.

Safety

VMS itself can help in search and rescue (SAR), especially when the SAR organization participates in the Global Maritime Distress Safety System(GMDSS). Some VMS have built-in Emergency Position-Indicating Radio Beacons (EPIRB), or SART capability although a dedicated VMS unit may not be able to have an emergency beacon that automatically floats to the surface and starts transmitting when it detects it is in salt water. At the very least, the SAR agency can get a last reported location of the vessel, and perhaps its course, from the FMC.

Enforcement

VMS obviously is part of fisheries enforcement, but, along with other systems, it can be part of overall sea surveillance. When a radar or other sensor detects a given vessel, VMS can tell the center that monitors the radar whether the radar target is a known fishing vessel. There may be correlation between AIS/VTS and VMS.

Technologies and Components

Any vessel tracking system, including VMS, requires technology on the vessel, ashore, and communications between them. In addition, there may be additional communications from the Fisheries Management Center (FMC) of the vessel's country of registry, and regional or national FMCs of the waters in which the vessel is fishing.

Functions Aboard the Vessel

The most basic function of a VMS is to determine the vessel's location at a given time, and periodically transmit this information, to a monitoring station ashore. Different VMS systems use different communication technologies, including AIS, Inmarsat, Iridium and Argos depending on the functionality required by the particular VMS system.

VMS components on the vessel sometimes are called VMS, or sometimes Automatic Location Communicators (ALC). These minimally include a GPS antenna and receiver, a computer (which may be embedded or user-supplied), and a transmitter and antenna appropriate for the communications that links the vessel to the flag center.

In practice, many of the VMS components also have applicability, along with non-VMS marine electronics, to a wide range of functions aboard a fishing vessel. These include navigation, finding fish, collision avoidance, routine voice and email communications, etc.

Selecting a VMS system is most dependent on what vendors and models have been approved by the fishing vessel's state of registry and the functionality requirements. Normally an authority will specify specific approved equipment to ensure end to end system integrity and service level meeting the specific requirements applicable to the vessel type. For example, some systems require a user interface on the vessel, whilst others will have a simple black box transceiver with no user interface.

VMS software and devices for the fishing vessel include:

- Absolute Software,
- AMS,
- Argonet-vms CLS,
- AST Ltd,
- Globavista/Bluefinger/Thales,
- BlueTraker,
- Boatracs,
- CLS America - Thorium VMS,
- Faria,
- Free Port - Eye from a sky,
- Fulcrum Maritime Systems - Vessel Tracking, Fleet Tracking, Fisheries Monitoring,
- Honeywell,
- Monicap,
- Skymate,
- SRT Marine System Solutions,
- Thrane & Thrane,
- vTrack VMS,
- MetOcean Telematics iTrac VMS,
- GMV.

Communications

VMS units principally rely on global navigation satellite systems (GNSS) such as GPS for position and time information. LORAN may be a backup or complementary technology. These transceivers transmit data to monitoring systems generally using a variety of communication technologies including terrestrial & satellite AIS and conventional satellite systems from Inmarsat, Iridium, Argos, ORBCOMM or Qualcomm. Increasingly nations are implementing a mixture of technologies

with the largest vessels over 60 tons being required to use the expensive traditional satellites and smaller vessels to use AIS.

- Operated by Inmarsat plc, founded by governments but now commercial, Inmarsat has a constellation of geosynchronous communications satellites.
- Iridium uses a constellation of 66 Low Earth Orbit satellites to provide complete global coverage (including all ocean regions and both poles) with real time coverage.
- Automatic Identification System - AIS is an IMO supported technology- provides low cost dual terrestrial and long range satellite maritime data communications. It offers a base level of VMS functionality with global coverage at the lowest cost.
- Argos uses Low Earth Orbit European and US satellites in polar orbit, which is an especially appropriate orbit for vessels operating in high latitudes.
- Skymate uses Orbcomm LEO satellites, which is optimized for machine-to-machine communications, potentially at lower cost than voice-capable satellite systems. They operate in the VHF and UHF bands, and have demonstrated an AIS capability.
- Qualcomm provides access to the Iridium satellite systems.
- BlueTraker uses both GPRS and Iridium constellation to provide the biggest flexibility and the lowest communication costs. The BlueTraker is a stand-alone device fully integrated including the antennas, the communication modules and a back-up battery. It is also e-logbook ready.
- iTrac is certified by the Canadian Department of Fisheries and Oceans, has integrated GPS and uses the Iridium satellite system network to provide communication as well as e-mail access.

Fisheries Management Center of the Vessel's Nation

Software at the fisheries management organization looks for several pieces of information:

- Location vis-a-vis restricted area,
- Time at sea,
- Time in restricted area, possibly separating fishing and transit time by speed.

A restricted area may be closed for all purposes, open to transit, open to fishing for a specific period, or open to designated vessels only. Vessel speed is often the way its status is determined in lieu of direct observation. Some VMS directly report speed, or speed can be calculated by FMS software based on the time stamps of different position reports. A rule of thumb in scallop fisheries, for example, is that the vessel cannot be dredging for scallops if its speed is greater than 5 knots (9 km/h).

FMC software can note the time a vessel leaves and returns to port, and the time it is inside or outside designated areas. There may be restrictions on trip length, time in an area, etc., which can be calculated directly from VMS data. Other observations may require correlation of catch reports with the vessel's presence in given areas. Presence in other areas may require an onboard observer.

Individual regional, national, and international FMCs have different levels of software intelligence, which can detect patterns of interest to SAR, fisheries management, or law enforcement.

Fisheries Monitoring Center of the Waters being Fished

Countries with registered fishing vessels that employ VMS generally agree to set up a Fisheries Monitoring Center (FMC), which has a data network connection to the FMCs of other states as well as other maritime stakeholders within the country. This flag state principle requires all vessels, registered in a given state to transmit their positions automatically to that state's FMC. When the vessel enters the waters of a different state, the home FMC must forward the report of the vessel's entry into those waters to the foreign state FMC. Until the vessel leaves the foreign state's coastal area, the home FMC must forward to the foreign state FMC the position, speed, and course reports at least every two hours. Exchange of data between VMS servers within the EU is regulated by the European Commission to be formatted according to the NAF FORMAT, originally devised by NAFO and subsequently adopted in a slightly different format by NEAFC. VMS servers outside of the EU may optionally also use the NAF format due to its widespread use within the EU. Data is usually transmitted using HTTPS protocol either by an HTTP Post or an HTTP Get request. Other protocols such as X.25 have historically been used but are in decline.

The NAF format is being replaced by FLUX, which defines messages and forwarding rules for many messages related to various business processes associated with VMS operation.

If position reports unexpectedly cease from a vessel, the FMC for the ocean area from which the last signal was received must attempt to contact the vessel or the flag state FMC without delay. Since VMS reports are sent automatically, it is possible that there is nothing wrong with the vessel itself, only the VMS. A full search and rescue (SAR) operation should not be launched simply because a VMS report does not arrive, although it is reasonable to alert sea surveillance assets, such as radars, that might be able to find the vessel. Fishing vessel crews should check the VMS at reasonable intervals, and confirm it is working.

While the procedure will vary with the jurisdiction, if an at-sea vessel finds their VMS is not working and they cannot fix it, they may be able to contact the FMC and get permission to continue the voyage. If they do get such authorization, they may get an inspection when they return to port. The FMC may also order them back to port. It is unlikely they will be allowed to leave port again without the VMS being repaired, so that they may need 24/7 VMS technical services at their home port.

Position reports received by the FMC should be forwarded automatically to the FMC of the vessel's registry. FMCs and other organizations, such as SAR and research, which receive VMS data must comply with confidentiality agreements. All recipients of data are also in accordance with agreements obliged to handle the data they receive in a responsible manner.

Catch Documentation Scheme

Catch reports are not themselves part of VMS, but often will be correlated with VMS data as part of an overall fisheries MCS program.

VTS and GMDSS

These are specific safety and traffic management systems which generally do not include specific fishery functionality as required by VMS.

AIS

AIS is often confused as a system comparable to VMS or VTS. AIS is a communications technology which is normally used within VTS and VMS applications. AIS is typically used on VMS systems deployed on smaller fishing vessels under 60 tons. The implementation of AIS as part of these systems is also often customized to include encryption and other functionality.

FISHERY RESOURCES MONITORING SYSTEM

The Fishery Resources Monitoring System (FIRMS) aims to systematically assemble comprehensive and reliable information on fisheries and fishery resources from the national, regional and global levels.

FIRMS was established in February 2004 to respond to the need for such fisheries information. It draws together a unified partnership of international organizations, regional fishery bodies and, in the future, national scientific institutes, collaborating within a formal agreement to report and share high-quality information on the global monitoring and management of marine fishery resources. FIRMS aims to provide decision-makers with the best and most authoritative information possible in order to develop effective fisheries policies in accordance with the Code of Conduct for Responsible Fisheries. To further the effectiveness of fisheries information management and dissemination, FIRMS also participates in the development and promotion of agreed global standards.

Born in parallel with the Strategy for Improving Information on Status and Trends of Capture Fisheries (Strategy-STF), FIRMS addresses at least five of the nine required actions of the Strategy.

The Information Backbone: Inventory of Stocks and Fisheries

FIRMS works as a monitoring system based on an inventory of stocks and fisheries conducted under the general framework of the 2003 FAO Strategy-STF which has been endorsed by the United Nations General Assembly. The inventory of marine resources monitored by FIRMS partners ranges from broad regional resources to more local resources or stocks. Each partner agrees to regularly provide relevant information on the exploitation, assessment, management and status and trends for species and fishing areas for which they have primary responsibility. In this way, FIRMS contains a unique and comprehensive vision of the current state of the world's fisheries. As the partnership grows, so will the reporting on marine resources and fisheries of the global inventory conducted under Strategy-STF.

Information is presented in synthesized Fact Sheets and State of Resources Summaries. These include images, maps of geographical distribution, general biological and habitat characteristics, scientific assessment results, management considerations and status and trends statements. Every Fact Sheet is sourced with relevant links thus allowing for transparent traceability, cites related references and, where available, mentions noteworthy observations from other sources. These fact sheets may be published in each of the five FAO official languages.

The FIRMS underlying system enables to maintain the history of fish stock assessments: who did what, where, when and how, and what were the results? Keeping a master record on individual stock assessments has significant value for a wide range of purposes:

- To facilitate retrospective analyses of stock assessments for the purpose of improving them.
- To provide a structured repository of information on assessment methods employed.
- To increase transparency.
- To allow access to other agencies' information in database format.

The relatively long time frames that are required to reach fishery sustainability for a stock rebuilding programme means that time series of information must be kept. These time series include data and estimates derived from analyses, as well as the methods used in obtaining the estimates.

The Operation: Shared Concepts and Standards

To ensure consistency and a common understanding of the information shared, FIRMS has established core concepts, definitions and data modelling. Partners have agreed on protocols and standards in their reporting, which enhance the overall authoritative value and quality of the information shared in FIRMS. FIRMS is powered by the FAO Fisheries Global Information System (FIGIS) and benefits from its content management system and information exchange protocol features. Information provided by partners through streamlined protocols is published on the Web for general access. FIRMS provides partners with the appropriate tools and training to ensure controlled dissemination of high-quality and updated information. Partners are committed to providing the best scientific evidence on the status and trends of fishery resources and fisheries, and to document the quality of their data contributions. For each collection of fact sheets, the FIRMS Data Quality Assurance statements describe the set of criteria applied, enabling users to evaluate FIRMS information content.

FIRMS Road Map and the STF Framework

In the long run, such recommendation for expanding the partnership will allow FIRMS to achieve full global coverage. By involving new partner contributors, particularly at national level, the effective gap between monitored and unmonitored resources and fisheries will be highlighted. This will assist policy-makers in identifying new areas for action and priority. It will also directly support both international and national plans of action for responsible fisheries, as well as bolstering actions needed to achieve the UN Millennium Development Goals. Although the FIRMS Secretariat and system maintenance is part of the FAO Regular Programme, the initial development phase has also been supported by donors (France with an APO, and Norway and Japan through the Fishcode-STF project). Strengthening and expanding the partnership will require locating additional financial resources to fund necessary conferences, workshops and growing maintenance activities and to keep pace with technological improvements. Where there are no FIRMS partners, FAO's short term move towards this long term goal is to develop regional or national inventories of marine resources and fisheries within the Strategy-STF framework, on a region by region basis.

FISHERIES SURVEILLANCE AND ENFORCEMENT

Effective management of coral reef fisheries cannot succeed without an adequate legal framework, effective law enforcement, and compliance efforts. However, inquiries with environmental and fisheries authorities around the world reveal a reality that is common in every surveyed country: there are not sufficient vessels and personnel for enforcement, and when vessels, personnel, and equipment do exist, few vessels are operative due to the lack of funds for spare parts, fuel, or routine maintenance operations. Furthermore, when patrols are carried out and poachers apprehended, the individuals who violate the law are rarely fined due to outdated laws, corruption, or non-existent judicial follow-up.

In above figure, local patrol staff of the Kofiau Field Station encounters a non-local fisherman operating illegally within the Marine Protected Area. The fisherman's catch is noted, and he is instructed as to the regulations in the area. Kofiau, part of the Raja Ampat Islands of Indonesia, is located in the Coral Triangle, an area containing what may be the richest variety of marine species and corals in the world. The people of Indonesia's Raja Ampat islands rely on the sea as their most important source of food and income.

Fishery management strategies such as the establishment of Marine Protected Areas (MPAs) or the closing of fishing areas or seasons often require the establishment of a framework where authorities, the private sector, local communities, NGOs, academic institutions, and other stakeholders agree to collective action. Establishing such a framework and enforcing respect for the law are the cornerstones of a good governance program.

An effective law enforcement system should dissuade potential lawbreakers from committing illegal activities, since it will ensure that the consequences/risks associated with apprehension outweigh economic gain. There are 5 key components in the development and implementation of an effective surveillance and enforcement system:

- Surveillance and Interception: One must first identify the most cost-effective suite of sensors for detection in a given area, and then use the information supplied by these to respond and intercept the illegal activity. The response depends on institutional or community capacity i.e. available vessels and staff, fuel, protocols, etc.
- Systematic Training: The regulations, systems, and tools are only as useful as those who are trained to operate and maintain them.
- Prosecution and Sanction: It is not worth investing millions of dollars in surveillance systems if there are no repercussions. Administrative or criminal prosecution and sanctions are necessary to ensure compliance.
- Sustainable Finance: Enforcement systems cost money. Funding for establishment and long-term operations must be identified and secured.
- Education and Outreach: It is critical to foster community buy-in as well as to inform stakeholders of rules, regulations, and sanctions.

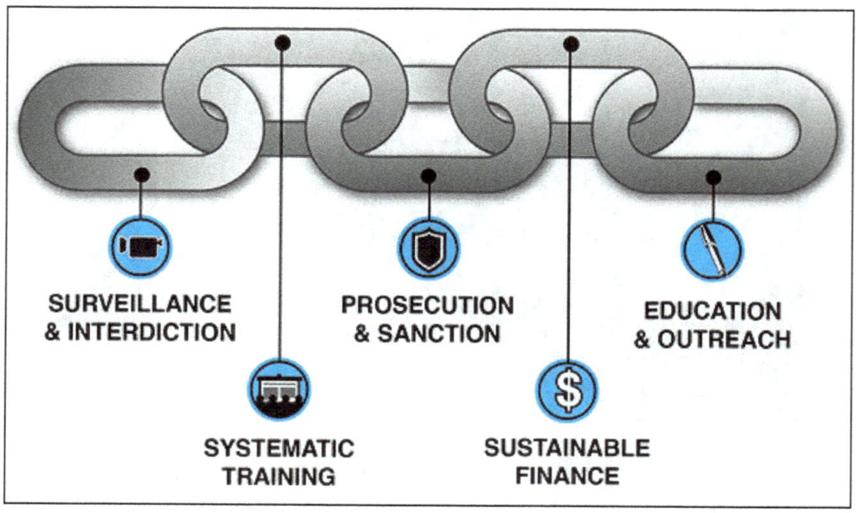

Surveillance and Interdiction

A surveillance system should include the most cost-effective suite of sensors for detection that provides information for appropriate response and for successful interdiction in a given area. Typically there are two types of systems that are used for surveillance: collaborative and non-collaborative.

Collaborative Surveillance Systems require active location transceivers on-board vessels. There must be a law that mandates the use of transceivers and dictates penalties for deactivation. Fishers must be part of the process for the implementation of this type of system. There are two main types of collaborative technologies:

- An Automatic Identification Systems (AIS) is an on-board vessel identification system that operates over marine VHF (Very High Frequency) and transmits vessel information like ship name, course, speed, and precise location in coastal and inland waters. AIS usually works in continuous mode and the system coverage range is similar to other VHF applications. The information is sent to shore base stations or vessel traffic service centers to monitor the movement of vessels.

- Vessel Monitoring Systems (VMS) is also an on-board vessel system that uses satellites to collect and transmit data on vessel name, location, course, etc.

Collaborative System Pros

- AIS provide real time monitoring of small and fast vessels.
- VMS provide location information over large areas and open ocean. VMS transmit an encrypted message (not open to the public, but directly to authorities).
- The capital and operating costs of both AIS and VMS are relatively low (costs can range from $50,000 for local confined system to $5M+ for a national system of about 8,000+ vessels) and both systems are being deployed in the developing world.

Collaborative System Cons

- VMS provide location information every 1–6 hours, which is not appropriate for artisanal vessels. In addition, a transceiver costs $800–$1,300 and there is a monthly cost associated with the service ($20–$60, depending on signal frequency).
- AIS also require the purchase of a transceiver. There are no recurring costs for the fisher when shore-based AIS stations exist; however, with the advent of AIS migration to satellites, recurring costs could also become a factor.
- Collaborative systems require a law to mandate their use and to provide incentives for adoption. For example, in Ecuador the use of transceivers conditions access to fuel subsidies.
- Transceivers can be deactivated and require stiff penalties for violations. In Mexico, there is a law that requires the installation of VMS transceivers on commercial fishing vessels, but the law does not specify penalties for transceiver deactivation. This loophole allows fishers to deactivate their transceivers if they do not want to be detected.

Non-Collaborative Systems do not require transceivers or the participation of stakeholders in the process. These surveillance systems detect vessels in a specific geographic area. There are several types of non-collaborative surveillance systems, such as visual, radar, optical (images taken from satellite, manned or unmanned aerial vehicles), and/or infrared cameras. These sensors are either located at strategic sites on the coastline and/or mounted upon mobile platforms, such as patrol vessels or drones. The monitoring of radio communication is another non-collaborative option for identifying suspicious activity.

Non-collaborative Systems Pros

- Radars are ideal for the detection of medium to large vessels up to 30 nautical miles (nm).

- Optical, video, and infrared cameras are ideal for the detection of vessels up to 10 nm.
- Beyond that range, they can be mounted on drones for an additional 20–30 nm of coverage.
- Radio monitoring can be a cost-effective way to detect suspicious activities.
- Visual observation by officers and stakeholders is limited to 8 nm and appropriate for community-based enforcement systems. They are the least expensive and are effective for near-shore contexts.

Non-collaborative System Cons

- Radars do not perform well beyond 6 nautical miles from the coast in the detection of small wooden or fiberglass vessels.
- Optical, video, and infrared cameras are limited to 10 nm.
- Infrared cameras are expensive.
- Drones are quite expensive, increasing in cost as their autonomy range increases beyond 5 nm.
- Radar monitoring requires dedicated and trained personnel.
- Visual observation by officers and stakeholders is typically limited to 8 nm.

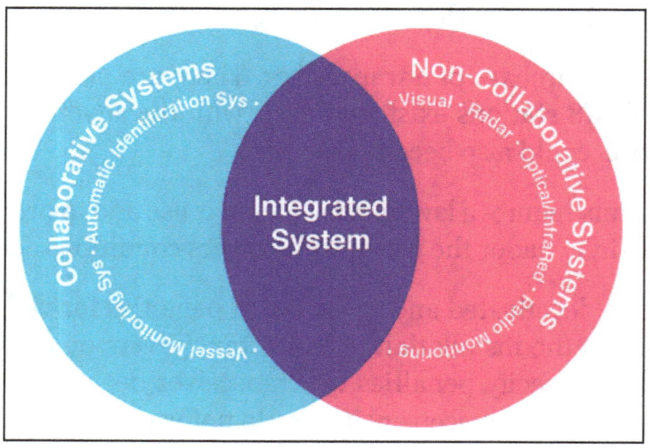

Intergrated system.

Integrated Surveillance Systems Approach

No one sensor can provide full coverage, so surveillance systems are often designed using a variety of sensors in an integrated system. To determine the necessary technology for designing a surveillance system, it is critical to consider detection distance, target size, and types of materials used for vessels (wood, fiberglass, or aluminum). For example, radars or high power cameras can be used in combination with VMS transceivers and AIS. The radars or cameras can be placed near productive fishing grounds to detect vessels that have deactivated their VMS transceivers and are fishing in a prohibited area. See the figure below for details on types of collaborative and non-collaborative surveillance technologies and their respective coverage ranges.

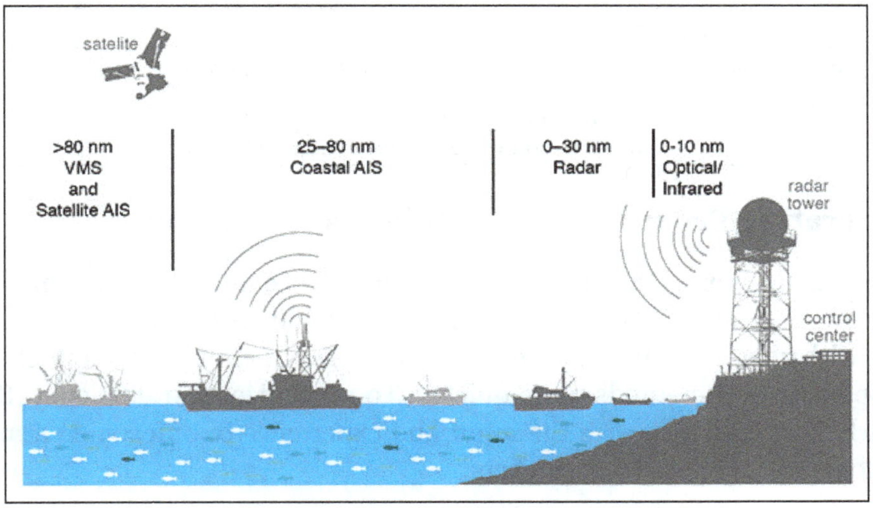

An Integrated Surveillance System.

Systematic Training

A comprehensive training program on the topics in the table below is required to strengthen the professional capacity of management and enforcement teams.

Subject	Description
Surveillance, Interdiction and Boarding	• Operational planning and preparation • Of visual and electronic sensors for patrols • Protocols: inspections, required documents, what to check and look for, documenting the inspection. Training should include prosecutors. • And confronting suspicious crews • Scene protocols. Collecting and handling of evidence • Operational report. Items and information that should go into a report
International Maritime Organization (IMO) Basic Training	• Aid • At sea • Fighting
Operational Planning and Administration of Control Center	• Control center functions, including risk assessment, use of assets, reports, communication protocols, surveillance and documentation protocols • Telecommunication lines and collaborative procedures with coastguards • Situational assessment and real-time reports • Reading and using nautical charts • Reading and using land maps • Search and rescue • Providing first-aid services in the field • Considerations for personal security during patrols and boarding

Subject	Description
Basic and Advanced Courses on Outboard Motor Maintenance	• A basic outboard motor maintenance course provided by the manufacturer • One or two rangers should be assigned to attend the advanced course on outboard motor maintenance and critical repair.

Standard Operating Procedures that Reinforce Training

Standard Operating Procedures (SOPs) will ensure that enforcement efforts are implemented effectively, that they guide daily tasks and institutionalize and maintain professional standards. SOPs, in combination with good reporting and feedback strategies, help new personnel learn appropriate actions, responses, and methods more quickly. To maximize their efficiency, SOPs should be updated regularly in accordance with the input and experience of officers. At a minimum, SOPs should be developed for the following subjects:

Control Center

The control center should be staffed by at least one officer at all times and ideally be operational 24 hours a day, 7 days a week. The operating officers communicate all infractions and events to the control center supervisor. The center is responsible for monitoring information concerning vessels near the monitored area and those entering and leaving the monitored area. Main responsibilities of the center include:

- Directing and ensuring communications between officers, vessels, and managers, as well as with other agencies.
- Coordinating active operations. The center also implements interceptions (or interdictions) and sends backup as needed.
- Maintenance of all archives including user manuals and SOPs. Communication with external agencies and managing confidential information.
- Maintenance of technology and resource inspections.Knowing that the personnel profile fits the functional needs of the different posts.

Patrol

- Pre-departure checks: verify that all the bridge gauges and indicators are operating, test the speed control and guidance system, prepare underway logs, personal equipment, navigation equipment – radar, GPS, echo sounder, charts, DF, check ship's logbook open and duly noted for the commencement of the patrol, and raise international fisheries pennant to show that you are on a fisheries patrol.
- Checks of other equipment: obtain machine report and verify that portable radios are functioning; check that boarding equipment such as life jackets, binoculars, rifle, hand guns, flares, boarding flags, net gauge, or other fishing regulation measuring equipment is on board.
- Submission of a pre-departure report to the operational director or supervisor.
- Establishment of patrol strategies, such as multiple boat patrol, patrol with a cross-search leader, barrier patrol, radar patrol, and patrol with searchlights.

Boarding

Boarding inspections are subject to national and international maritime control and interdiction proceduresand must take into account a range of potential activities, from fishing violations to greater crimes such as drug trafficking, piracy, contraband, or murder. Boarding inspections may be met with an armed and hostile response. Therefore, the highest level of precaution for personnel and vessels is advisable. At a minimum, boarding SOPs should:

- Determine if patrols will be performed undercover.
- Determine the distance and speed of vessels to be intercepted and detained.
- Delineate minimum training requirements for personnel in the inspection of different types of vessels and their associated risks.
- Establish protocols for the chain of command, control, and abnormal situation assessment (ex: the escalation of a detected crime).
- Establish communications protocols to keep constant communication with the control center (ex: perform periodic checks every 15 minutes).
- Set restrictions on the use of cell phones or personal cameras while performing a boarding inspection (these can put the security/success of the operation at risk). Allow only the team leader to use them.

Prosecution and Sanction

Enforcement systems require effective criminal, civil, and/or administrative sanctions and punitive actions. Prosecution is different in every country, but the sanctioning process for environmental violations is typically very slow. Effective compliance systems are important in order to avoid delays and lost trials, which ultimately cause economic losses for the state in terms of wasted patrol resources and loss of natural capital. While the legal framework is unique in each country, the following types of sanctions should be considered in the development of enforcement systems:

Criminal/Civil Sanctions: The following actions are recommended in order to improve judicial proceedings:

- Establish a standardized boarding report format with recommendations from the relevant legal authority's office.
- Train officers to fill in reports according to this format.
- Formalize official relations between officers patrolling the area and their provincial and/or federal counterparts.
- Carry out training workshops for judges, attorneys, and lawyers once a year on marine and fisheries-related rules and regulations.
- Assign additional lawyers from NGOs or support agencies to follow up on environmental marine violations or crimes.

- Set up private prosecutions for major cases using external lawyers.

Administrative Sanctions: In order to expedite the sanctioning process, where possible administrative sanctions should be carried out at the local level. The severity of measures should correspond to the seriousness of the violation. Non-economic sanctions should also be considered, for example:

- Vessel detention for a limited time;
- Restriction of sailing authorization permits;
- Seizure of fishing gear;
- Temporary suspension of permits for ships, crew members, or ship-owners;
- Revocation of operating licenses for ships, ship-owners, agents, maritime personnel, or fishers.

Sustainable Finance

Effective enforcement systems require funding. However, most marine resource enforcement efforts receive limited central government support, leaving administrators with very little resources to enforce laws. User fees and increased cost-effectiveness of operations are two common strategies to sustain enforcement efforts.

User Fees: The governments of Ecuador and Palau have initiated protected area taxes to directly support conservation efforts. The green tax in Palau, for example, raises over $5M annually, while the Galapagos entrance fee raises approximately $1M for enforcement in the marine reserve. These revenue-generating schemes are politically feasible, appeal to tourists who want to help protect natural and cultural resources, and can provide critical funding for enforcement. Unfortunately, the legal frameworks in many countries do not permit the creation and administration of user fees. As a result, some resource management agencies and MPAs have developed co-management agreements with NGOs, tourism operators, or communities to increase funds and capacity for enforcement efforts. For example, the NGO WildAid manages external donations from the tourism industry for the Galapagos National Park Service to fund critical spare parts and patrol vessel maintenance.

Decreasing Cost of Operations: Enforcement programs should also look to increase the cost-effectiveness of operations. In programs evaluated by WildAid, typically 70% of maritime enforcement costs are comprised of personnel salaries and fuel. At least two strategies can support cost-effective enforcement operations that reduce fixed costs while ensuring that the area to be monitored can still be covered:

- Performance-Driven Acquisitions: It is important to conduct a comprehensive analysis of the fishers/area users, violators, the geographic area, and institutional capacity when designing an enforcement system. Appropriate staffing, surveillance equipment, number and type of fishing vessels, and outboard motor size are some of the critical variables to consider in the design of an enforcement system.

What not to do: Many sites make the mistake of having a donor or a sales representative prescribe the technology or the enforcement system design, which often results in a system that is overly sophisticated and costly for the specific area. There are many examples of donated vessels that end up costing more to maintain in the long run than they are worth to enforcement efforts.

- Use-Appropriate Technology: Enforcement strategies that combine technology with patrols are most cost-effective. As most sites possess a limited maritime staff, systems can be designed that combine the use of surveillance sensors with the strategic placement of buoys for the mooring of patrol vessels that remain in close radio communication with the control center officer and respond only when a potential violation occurs. Although this strategy does not eliminate the need for some patrols, it does reduce the amount of fuel used, as the vessels can sit idle at moorings for extended periods of time.

  What not to do: Carry out constant patrols that cost staff time, fuel, and wear and tear on vehicles and vessels.

Fisheries Management

5 CHAPTER

Fisheries management is the activity of protecting fishery resources in order to facilitate sustainable exploitation. It can be broadly categorized into conventional fishery management and ecosystem-based fisheries management. This chapter discusses in detail these types of fisheries management as well as the different fish diseases and parasites.

Fisheries management refers to a governmental system of management for the protection and sustainable development of biological resources. Currently, fisheries management focuses more on limiting catch rather than proactive marine resource recovery measures, such as aquaculture. Compared to agricultural studies and technology, marine science and technology are still at an early stage of development and the development of fisheries management will require further research and public support. Effective fisheries management can contribute to the sustainable development of marine resources.

The control of fisheries production has been exercised in many places around the world for hundreds of years. For example, the Māori people, residents of New Zealand for about the last 700 years, had strict rules in their traditional fishing activities about not taking more than could be eaten and throwing back the first fish caught (as an offering to Tangaroa, god of the sea). Another longstanding example is the North Norwegian fishery off the Lofoten islands, where a law has existed for more than 200 years to control fishing activity; in this case, the increase in the number of fishermen caused problems that necessitated management. To avoid gear collision, gillnetters and longliners are separated and not allowed to fish in the same grounds south of Lofoten. Similar kinds of fisheries management have been practiced in many places of the world, which often involve aspects such as restrictions on fishing seasons, location, and species.

Fishery management in the pre-modern era has traditionally been organized by local fishing communities. Their practices have often been connected with religious beliefs, rituals, and customs.

These forms of community based fishery management, however, encountered various challenges in the modern era as each community, and their marine resources, became a part of national or global economy and fisheries were managed primarily from the perspective of economic growth, with little to no attention given to ecological sustainability.

Governmental resource protection-based fisheries management is a relatively new idea, first developed for the North European fisheries after the first Overfishing Conference held in London in 1936. In 1957 the British fisheries researchers, Ray Beverton and Sidney Holt, published a seminal work on North Sea commercial species fisheries dynamics. The work was later (in the 1960s) used as a theoretical platform for the new management schemes set up in North European countries.

After some years away from the field of fisheries management, Ray Beverton reassessed his earlier work and in a paper given at the first World Fisheries Congress in Athens in 1992, he criticized some of the concepts that he had earlier laid out in "The Dynamics of Exploited Fish Populations" and expressed concern at the way his and Sydney Holt's work has been misinterpreted and misused by so many fishery biologists and managers during the previous 30 years. Nevertheless, the institutional foundation for modern fishery management had been laid.

Objectives

There are various political goals in fisheries management; however, these objectives can conflict with one another. Typical political objectives include:

- Maximize sustainable biomass yield.
- Maximize sustainable economic yield.
- Secure and increase employment in certain regions.
- Secure protein production and food supply.
- Increase income from export.
- Biological and economic yield.

Rules

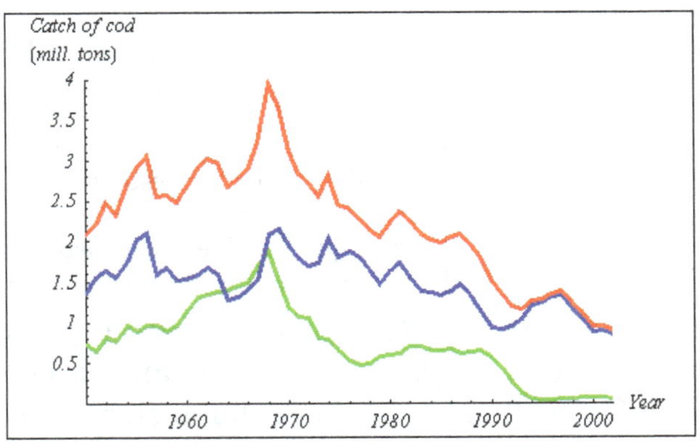

Catch of Atlantic cod 1950-2002 (FAO).

International agreements are required in order to regulate fisheries taking place in areas outside national control. The desire for agreement on this and other maritime issues led to the three conferences on the Law of the Sea, and ultimately to the treaty known as the United Nations Convention on the Law of the Sea (UNCLOS). Concepts such as exclusive economic zones (EEZ, extending 200 nautical miles (370 km) from the nation's coasts; if a continental shelf continues, EEZ is extended beyond 200 nautical miles) allocate certain sovereign rights and responsibilities for resource management to individual countries.

There are a number of situations that need additional intergovernmental coordination. For example, in the Mediterranean Sea and other relatively narrow bodies of water, EEZ of 200 nautical miles are irrelevant, yet there are international waters beyond the 12-nautical-mile (22 km) line of coastal sovereignty. International agreements, therefore, must be worked out for fishery management in the international waters of the narrow sea.

There are also issues with straddling fish stocks. Straddling stocks are fish stocks that migrate through, or occur in, more than one economic exclusion zone. Here sovereign responsibility must be worked out in collaboration with neighbouring coastal states and fishing entities. Usually this is done through the medium of an intergovernmental regional organisation set up for the purpose of coordinating the management of that stock.

UNCLOS does not prescribe precisely how fisheries that occur solely in international waters should be managed, and there are several new fisheries (such as high seas bottom trawling fisheries) that are not yet subject to international agreement across their entire range. Both of these issues came to a head within the United Nations in 2004 and the UN General Assembly issued a resolution on Fisheries in November 2004 which set the scene for the further development of international fisheries management law.

Fisheries objectives need to be expressed in concrete management rules. In most countries the management rules today should be based on the internationally agreed, albeit non-binding, standard Code of Conduct for Responsible Fisheries, agreed at an FAO (the UN Food and Agriculture Organization) session in 1995. The precautionary approach prescribed here is also implemented in concrete management rules as minimum spawning biomass, maximum fishing mortality rates, and so on.

Mechanisms

When it comes to controlling the activities of individual fishers or fishing operations (vessels or companies), available management means can be sorted into four categories:

Taxation on input; vessel licensing	Taxation on output; restrictions on catching techniques
Limited entry control	Catch quota and technical regulation

The top row represents indirect methods while the bottom row represents direct methods of regulation. vessel monitoring systems, patrol vessels and aircraft, and observers aboard fishing vessels are examples of direct regulatory methods. The left column shows input controls and the right column output controls.

Many countries have set up Ministries and Government Departments, named "Ministry of Fisheries" or similar, controlling aspects of fisheries within their exclusive economic zones.

Performance

According to a 2008 UN report, the world's fishing fleets are losing $50 billion USD each year through depleted stocks and poor fisheries management. The report, produced jointly by the World Bank and the UN Food and Agriculture Organization (FAO), asserts that half the world's fishing fleet could be scrapped with no change in catch. In addition, the biomass of global fish stocks have been allowed to run down to the point where it is no longer possible to catch the amount of fish that could be caught.

Elderly Maternal Fish

Traditional management practices aim to reduce the number of old, slow-growing fish, leaving more room and resources for younger, faster-growing fish. Most marine fish produce huge numbers of eggs and larvae, and the assumption has been that the younger spawners will produce plenty of viable larvae.

However, recent research on rockfish shows that large, elderly females are far more important than younger fish in maintaining productive fisheries. The larvae produced by these older maternal fish grow faster, survive starvation better, and are much more likely to survive than the offspring of younger fish. Failure to account for the role of older fish in maintaining healthy populations may help explain recent collapses of some major US West Coast fisheries. Recovery of some stocks is expected to take decades. One way to prevent such collapses may be to establish marine reserves, where fishing is not allowed and fish populations age naturally.

Ecosystem Approaches

Large Aquaculture facility in Martin Bord, West Bosnia.

According to marine ecologist Chris Frid, the fishing industry has been keen to identify pollution and global warming as the causes of unprecedented low fish levels in recent years. But it is clear that overfishing has also altered the way the ecosystem works. "Everybody would like to see the rebuilding of fish stocks and this can only be achieved if we understand all of the influences, human and natural, on fish dynamics." Frid adds: "Fish communities can be altered in a number of ways, for example they can decrease if particular sized individuals of a species are targeted, as this affects predator and prey dynamics. Fishing, however, is not the sole perpetrator of changes to marine life—pollution is another example. No one factor operates in isolation and components of the ecosystem respond differently to each individual factor.

The traditional approach to fisheries science and management has been to focus on a single species. This can be contrasted with the ecosystem-based approach. Ecosystem-based fishery concepts have existed for some years and have been implemented in some regions. In a recent (2007) effort to "stimulate much needed discussion" and "clarify the essential components" of ecosystem-based fisheries science, a group of scientists have offered the following "ten commandments for ecosystem-based fisheries scientists"

- Keep a perspective that is holistic, risk-adverse and adaptive.

- Maintain an "old growth" structure in fish populations, since big, old and fat female fish have been shown to be the best spawners, but are also susceptible to overfishing.
- Characterize and maintain the natural spatial structure of fish stocks, so that management boundaries match natural boundaries in the sea.
- Monitor and maintain seafloor habitats to make sure fish have food and shelter.
- Maintain resilient ecosystems that are able to withstand occasional shocks.
- Identify and maintain critical food-web connections, including predators and forage species.
- Adapt to ecosystem changes through time, both short-term and on longer cycles of decades or centuries, including global climate change.
- Account for evolutionary changes caused by fishing, which tends to remove large, older fish.
- Include the actions of humans and their social and economic systems in all ecological equations.

Aquaculture, including artificial hatching and fish farming, can be an important part of fishery management. However, aquaculture is still in its early stages of development. Complete life cycle farming, from hatching to spawning, is successful only for a few species for the mechanisms and conditions necessary for the vast majority of species are still unknown. The development of aquaculture will require further research and public support.

Fisheries Management Techniques

Fisheries resources must be managed to ensure that stocks are harvested at sustainable levels for the benefit of present and future generations. The fisheries resources of New South Wales are very diverse, and many different species are highly regarded by recreational fishers for sport and food, by the commercial sector for their monetary value, and by the community as fresh, quality seafood.

There has been a long held perception that because the commercial fishing sector catch a large quantity of fish it needs to be heavily regulated. Whilst this is certainly the case in some circumstances, there is an increasing recognition that an estimated one million recreational fishers who fish each year have a significant impact on our fisheries resources. Therefore, it is also important to manage the recreational sector.

Commercial Fisheries Management

Fisheries agencies throughout Australia have traditionally categorised fishery resources into distinct segments for ease of management. This is particularly necessary because of the large number of species taken for sale, and the range of gear types used by the commercial sector. Commercial fisheries have historically been defined by the method of capture (e.g. prawn trawl and fish trap), however, some of the more valuable fisheries with one target species are defined by the species itself (e.g. abalone and rock lobster).

Once each fishery is defined, policies and legislation can be developed to establish specific rules. These rules may then be included in a management plan for each fishery. Developing a fishery management plan involves extensive consultation with the commercial sector, the recreational sector and the community. In most cases, a specialised committee is established to develop a management plan acceptable to all.

Commercial Management Tools

There are two broad types of fishery management tools; input controls and output controls. Input controls limit the amount of effort commercial fishers put into their fishing activities, indirectly controlling the amount of fish caught. Input controls can include restrictions on the number of licences, the size and engine capacity of boats, the length and mesh size of nets, and the areas and times which can be worked. In some locations, commercial fishers are not permitted to work during weekends or public holidays.

A good example of how input controls work is the juvenile king prawn closure near the mouth of the Richmond River at Ballina. King prawns live in the estuaries as juveniles and migrate to the ocean during the young adult stage of their life cycle. This makes them vulnerable to capture at a small size by prawn trawlers. To protect the stocks of smaller prawns and to increase the value of the prawns sold, prawn boats are prohibited from trawling within an area bounded by a 2.2 nautical mile radius from the mouth of the Richmond River. This management measure also helps to reduce the capture of juvenile fish in prawn trawl nets.

Output controls, on the other hand, directly limit the amount of fish which can be taken from the water. The first step in implementing an output control management regime is setting a Total Allowable Catch (TAC) for the species. Once established, the TAC can be either fished on a competitive basis or divided up between the participants in the fishery so that all fishers have an individual quota. A TAC which is divided between the participants means that in any one year a commercial fisher is not able to catch more than their allocated weight of that species.

Output controls are generally regarded as good mechanisms to control the total catch in single species, high value fisheries which are targeted using a single gear type (such as abalone and lobster). However, thorough monitoring schemes such as daily catch logbooks are often required to ensure that individual quotas are not exceeded.

Recreational Fisheries Management

The goals of recreational fisheries management differ from those of commercial fisheries, as they have the broader aim of providing for the optimum use and recreational enjoyment of fisheries resources. Management arrangements are put in place both to control the total harvest, and to share the angling experience amongst fishers.

There are many restrictions on the types of recreational fishing gear that are allowed. Anglers are mainly restricted to line fishing methods with limited hook numbers, and a few small nets. These restrictions operate as input controls, as do closures on fishing in certain areas (e.g. marine reserves) and closed seasons (e.g. Murray cod spawning season).

Output controls for recreational fisheries take the form of daily allowable catches (bag limits) and size limits. Bag limits are set to limit the total harvest of certain species to sustainable levels, and to share the catch. In the absence of comprehensive scientific data, they are usually based on an amount of fish which is considered as socially acceptable to take in one day (e.g. enough to feed the average family). For example, anglers are only allowed to keep ten snapper per day.

Setting a minimum size limit is a management tool regularly used to ensure that most of the fish in the population are able to grow to breeding size and spawn at least once before they are caught. For example, the size limit on turban shells in NSW is 7.5 cm, a length at which 100% of the shells sampled were of reproductive age. This attempts to make sure that there will always be juvenile fish available in following years to grow and enter the adult population. Maximum size limits also apply to certain species because larger fish often breed more efficiently than smaller fish (eg. anglers may only keep two mulloway above 70 cm in length, per day).

Habitat Management

Fisheries management will only be effective in the long term if fish habitat is protected. The ever-increasing coastal population has resulted in increased land and water pollution, and greater pressure for coastal development. Much of the increased pressure affects areas near the edges of estuaries and bays, both of which are vital for the survival of juvenile fishes and crustaceans.

CONVENTIONAL FISHERY MANAGEMENT

Conventional fishery management seeks to maintain high, yet sustainable, yields by regulating the number or weight of fish caught, the size of fish caught, or the time and space (area) within which fishing is allowed. The intent in each case is to control fishing mortality rates. Conventional approaches to fishery management can be succinctly characterized by three main components: (1) an underlying fishery science and management paradigm, (2) a set of conventional management tools.

Fishery Paradigm

Fishery management relies on estimates of the population size of a target species to determine how many fish or what fraction of the population's biomass can be caught without damaging its reproductive potential. To make these determinations, management depends on a conceptual model of a fishery that makes three simplifying assumptions: (1) the fishing fleet targets and exploits a single-species stock, (2) the stock of interest is segregated temporally or spatially from other stocks, and (3) the individuals are perfectly mixed so that the effects of fishing are well spread over the whole stock. These assumptions, which are far from true in most situations, can have serious consequences for the effectiveness of fishery management.

Single-species, One Stock

Most management measures are directed at individual stocks of a single species and do not take into account species interactions, such as predator-prey relationships. A basic assumption of most models used to determine a catch level is that the catch rate a stock can sustain can be designated based upon the average productivity of the stock. Productivity, in turn, is presumed to depend primarily on the size of the adult stock. In this scenario, controlling adult stock size is the primary means of ensuring sustainability of the fishery. Furthermore, stocks are assumed to respond in a density-dependent manner and therefore are postulated to have maximum productivity at intermediate stock sizes. Thus, maintaining the stock size that allows maximum sustainable yield historically has been a major management goal, and fishing at a rate that produces MSY on average over good and bad years has been the target. In fact, fishing at the MSY level (a fixed exploitation rate policy) does not ensure constant catches in the future or a stable adult population size because of substantial variability in reproductive success and recruitment. It was recognized more than two decades ago that it is therefore too risky to set a constant quota for catch at MSY. In good years, fishers may prosper with MSY-based catches, but in years when the environment is less favorable and recruitment and productivity decline, the stock will diminish and MSY may quickly lead to overfishing. The fishing rate corresponding to MSY (F MSY) still remains a criterion in determining optimal yield, the regulatory target used to manage marine fish stocks. However, FMSY now is viewed by many as a threshold that should not be exceeded, rather than as a target at which to aim. More conservative quotas and exploitation rates are now recommended, due in part to recognition of our limited ability to estimate and implement F MSY or, for that matter, other target fishing mortalities via catch or effort control. Obtaining an estimate of MSY (or F MSY) usually requires fishing at levels that already exceed it.

Fish Stocks that are Temporally or Spatially Segregated

Although it is obvious that management must be tailored to individual species' life histories, the individual stock and single-species approaches to management are ineffective for multispecies or even mixed-stock associations, in which many different species or stocks with similar habitat and prey requirements overlap in their ranges. Good examples are reef fish off the southeastern United States and rockfish in the northern Pacific Ocean. Warm-temperate species such as gag (Mycteroperca microlepis), scamp (Mycteroperca phenax), and red hind grouper (Epinephelus guttatus), for instance, co-occur to such an extent that catch restrictions placed on one species in the complex typically result in increased regulatory discards while fishing for associated species. The same is true for the 83 species of rockfish managed as a complex off the Pacific west coast. Inevitably, regulatory discards will increase the mortality of the restricted species and threaten its recovery.

Individuals are Perfectly Mixed

A key element of the fishery management paradigm is the concept of a well-mixed stock. Migration patterns and more general spatial processes are fundamental components of fishery science. Knowledge of spatial processes serves to delineate management units, each viewed as a "dynamic pool" isolated from the rest. Most theory and management have been conceived for large-scale, commercial fisheries that target relatively mobile species (e.g., tuna, plaice, gadoids), for which dynamic pool assumptions may provide a reasonable simplification, at least at the scale of a fishing ground. In this conceptual model, because the effects of fishing are "diluted" in the pool, the use of spatially explicit approaches to manage each unit has been largely missing. This conceptual model has dominated marine fishery management. As a result, it has been applied indiscriminately to low-mobility species for which the paradigm is clearly inappropriate, such as rockfish (Sebastes spp.) and sedentary invertebrates. Fishing effort on relatively sedentary species preferentially targets the highest fish concentration and results in a mosaic of fishing mortalities and serial depletion of fishing grounds. Management by conventional means is complicated because the dynamics of the resource, fishing process, and monitoring are dominated by local processes that are often impossible or impractical to assess. In reef fisheries, for example, the relevant scale for assessment and management may correspond to a single reef.

Another tool to limit fishing effort is control of access to the fishing grounds. This strategy, known as "spatial management," can be applied as either temporary (seasonal or year-to-year) or permanent closure of portions of the fishing grounds. Spatial management can be used to control access to resources and probably has been practiced for centuries, but it has not played a central role in the management of marine fisheries. Its importance in the management of benthic shellfish is now accepted, as are novel management schemes involving area rotation. However, many more species could benefit from spatially explicit management, particularly those with a relatively stationary adult stage. A major constraint on implementation of spatial management is the lack of spatial catch data for many species and the complexity of spatially explicit stock assessment models.

Although spatially explicit components usually are missing, some existing fishery models do recognize that variables other than adult stock size affect productivity. However, these variables usually are treated as random variations beyond human control. Consequently, management focuses on regulating the size of the catch and the effort directed at obtaining it, while environmental factors are downplayed. Even when not controllable, persistent trends or variations in productivity driven by environmental conditions should be considered in policy evaluation. In addition, environmental variables that affect habitat quality and are directly affected by human activities are now addressed explicitly in fishery management plans and its emphasis on essential fish habitat (EFH). This focus on habitat has initiated a shift toward spatial considerations and designation of areas important for the productivity of economically important species. These designations, of essential fish habitat and habitat areas of particular concern, have bolstered interest in protected areas and supported the use of MPAs as a legitimate tool for fishery management.

Long-term, large-scale closures have been instituted as single-species refuges from fishing to promote rebuilding of depleted stocks. In many respects, they resemble permanent reserves, except that such areas may revert to their former status when restoration is attained. Long-term closures are becoming more common as a population-rebuilding tool (e.g., northern cod stock off Newfoundland), in which depleted populations may require many years to restore. Long-term closures might achieve some goals of reserves, although benefits may be transient if subsequent fishing mortality cannot be controlled.

Another form of spatial management is the use of rotating fishing areas. Here only a fraction of the fishing grounds is opened in a given season, the rest being closed for specified periods (often years) to promote growth of young animals and allow them to reach more valuable sizes. The area opened is rotated from year to year. This approach combines temporal and spatial closure to regulate fishing. In addition to rebuilding fish stocks, rotating closures may allow habitat and biological communities to recover from the effects of fishing, such as damage to bottom habitats by trawling and movement of "fixed" gear. However, recovery of habitat and biological communities may require closures on the order of 5 to 10 years .

Conventional Management Tools

Management based on the fishery paradigm above centers on measures that regulate fishing activities and the level of catch, rather than on measures that directly promote management of habitat or consideration of environmental variables affecting fish productivity. Generally, the goal is to manage exploited populations such that they are maintained at productive levels (close to or above MSY) that support a high yet sustained fishing yield and to require rebuilding plans when yields fall below a minimum stock size threshold.

Conventional management approaches to control exploitation rates fundamentally rely on placing limits on the amount or efficiency of fishing effort (effort-controlled fisheries) or on the total amount caught by specifying catch quotas and allocations (quota-based fisheries). In both cases, a target fishing rate is first specified, whether constant or variable in response to stock condition, based on analysis of historical experience with the fishery, on experience with fisheries for similar species, or on modeled responses of the fishery to simulated fishing mortality.

Effort Controls

There are many forms of effort controls, including restrictions on gear, vessels, time fished, and number of fishers. These are usually the first controls applied to a fishery to slow the rate of catch. Gear restrictions can include the type, amount, or dimensions of gear or specific features of the gear such as net mesh size, hook spacing on longlines, or configuration of fish traps. Vessel restrictions may include design, length, or engine horsepower. The number of fishers can be regulated by allocating licenses to either fishers or vessels. Time fished can be regulated through limiting the amount of time available for fishing through seasonal closures, "days-at-sea" restrictions, or specific days and hours when fishing is permitted.

To implement a target fishing rate by means of effort regulations, managers need a reliable estimate of the fishing mortality caused by each unit of fishing effort to be allowed, a parameter known as catchability. The estimate of catchability is then used to determine the amount of fishing effort (e.g., the number of total days at sea to be allowed) that is compatible with a chosen exploitation rate or fishing mortality so that:

$$F_{target} = \text{catchability} \times \text{effort}.$$

Catchability is generally assumed to remain constant as stock size varies. However, catchability may change—for instance, fishers may increase their efficiency when the stock declines—and this could result in overexploitation, a common reason for the failure of conventional management.

In addition to measures that attempt to regulate exploitation rate directly, other forms of effort control may be implemented to reduce fishing power, protect vulnerable life-history stages, or increase the market value of the fish. Temporal closures, for example, are frequently used to protect fishery resources at times when they are particularly vulnerable to fishing, such as when fish aggregate on spawning grounds. Also, temporal closures can be used to allocate fishing over the season in a manner that increases the value of landings. For example, fishing might be prohibited during parts of the year when the population is composed of small or poorly conditioned fish. Closure of the Dungeness crab (Cancer magister) fishery during the molting season when meat quality is poor provides one example. Broader application of temporal closures to protect whole communities or complexes of species and habitats is less common, but perhaps of greater relevance to developing marine reserves as an ecosystem approach to fishery management.

Catch Controls

A common approach to controlling fishing is to regulate the catch or the amount of fish landed. This is the favored method used to regulate fisheries in Alaska and along the west coast of the United States. Quota-based management relies on the ability to model relative trends in abundance over time and to estimate the absolute size of the exploitable stock. This information is used to set the total allowable catch (TAC) that meets a chosen target exploitation rate.

TACs or quotas are typically calculated as the product of a target exploitation rate μ and an estimate of current stock biomass $\mathrm{B\grave{I}}_{„t}$:

$$\mathrm{TAC}_t = \mu_t \mathrm{B\grave{I}}_{„t}.$$

In principle, TAC-based management can be a direct, efficient way to limit catches. However, the success of TAC-based systems depends on accurate estimates of stock abundance and biomass. Because the required level of accuracy is usually not available, the risk of overfishing may be high.

In many heavily exploited fisheries, both effort and catch controls are used to manage the fishery. In the Pacific halibut fishery, for example, a catch quota is used to control the exploitation rate, there is a minimum size limit on landed fish, and all fishing methods except setline gear are prohibited for the directed fishery. Seasonal closures are in place, which prevent the interception of fish from different regulatory areas when the fish migrate from feeding to spawning.

Both forms of regulation, catch and effort controls, have significant shortcomings because both depend on the quality of stock assessments. In turn, the accuracy and reliability of stock assessments depends on data that are frequently limited or unavailable. Conventional methods used to estimate stock abundance and current rates of fishing mortality require good historical catch statistics under significant levels of exploitation and indices of stock abundance that reliably show population trends. Numerous methods are used for stock assessment most are based on analyzing the way abundance changes in response to known catch levels. In addition to estimates of stock abundance or fishing mortality, conventional management depends on knowledge of what exploitation rates are adequate to derive "biological reference points" that designate both threshold and target levels of biomass and fishing rates. The sophistication of procedures used can lead to overconfidence in their ability to estimate abundances of stocks and their resilience to fishing pressure. This misplaced confidence contributed to the collapse of the Newfoundland cod fishery. Although stock assessments usually are done competently by fishery scientists, the statistical uncertainty associated with estimates and biological reference points can lead to failed management.

ECOSYSTEM-BASED FISHERIES MANAGEMENT

Ecosystem-based fisheries management is a holistic way of managing fisheries and marine resources by taking into account the entire ecosystem of the species being managed. The goal of ecosystem-based management is to maintain ecosystems in a healthy, productive, and resilient condition so they can provide the services humans want and need. The EBFM approach also can be applied in the management of protected and other trust marine species.

The traditional management strategy for fisheries and other living marine resources was to focus on one species, in isolation. For example, if a particular species' population was declining, fishery managers might reduce the annual catch limit the following year in an attempt to reduce overexploitation. However, fishing is only one variable affecting a species' population. Additional elements come into play, such as interactions with other species, the effects of environmental changes, or pollution and other stresses on habitat and water quality.

Benefits of Ecosystem-based Fisheries Management

EBFM is beneficial in decision-making, and improves our ability to predict the impact of those decisions. It is also cost-effective and designed to be adaptive. Specifically, EBFM:

- Facilitates trade-offs between different stakeholder priorities, balancing social and ecological needs.
- Forecasts pressures and impacts on both single and aggregated components of a marine ecosystem, and provides a better understanding of how ecosystems and their components respond to multiple stressors.
- Provides more stability of ecosystem level measures.

NOAA Fisheries has developed an agency-wide EBFM Policy, which outlines a set of principles to guide our actions and decisions over the long term. It directs continued progress toward development and implementation of EBFM approaches. It also ensures our commitment to incorporate EBFM into the agency's resource management decisions.

Integrated Ecosystem Assessments

Integrated ecosystem assessments are a way to better manage resources, and they provide a sound scientific basis for EBFM. These assessments provide a structure to assess ecosystem status relative to objectives of different groups (e.g., fishing, recreation, energy production, shipping, agriculture, forestry, food, clean water), account for the holistic impact of management decisions, and guide management evaluations.

IEAs consist of analyses that help resource managers make more informed and effective management decisions. For example, IEAs contain:

- Assessments of status and trends of the ecosystem condition, including ecosystem services.
- Assessments of activities or elements in an ecosystem that can stress it.
- Prediction of the future condition of the ecosystem under stress if no management action is taken.
- Prediction of the future condition of the stressed ecosystem under different management scenarios, and evaluation of the success of management actions in achieving the desired target conditions.

FISHERIES LAW

Fishery on Lake Tondano, Indonesia.

Fisheries law is an emerging and specialized area of law. Fisheries law is the study and analysis of different fisheries management approaches such as catch shares e.g. Individual Transferable Quotas; TURFs; and others. The study of fisheries law is important in order to craft policy guidelines that maximize sustainability and legal enforcement. This specific legal area is rarely taught at law schools around the world, which leaves a vacuum of advocacy and research. Fisheries law also takes into account international treaties and industry norms in order to analyze fisheries management regulations. In addition, fisheries law includes access to justice for small-scale fisheries and coastal and aboriginal communities and labor issues such as child labor laws, employment law, and family law.

Another important area of research covered in fisheries law is seafood safety. Each country, or region, around the world has a varying degree of seafood safety standards and regulations. These regulations can contain a large diversity of fisheries management schemes including quota or catch share systems. It is important to study seafood safety regulations around the world in order to craft policy guidelines from countries who have implemented effective schemes. Also, this body of research can identify areas of improvement for countries who have not yet been able to master efficient and effective seafood safety regulations.

Fisheries law also includes the study of aquaculture laws and regulations. Aquaculture, also known as aquafarming, is the farming of aquatic organisms, such as fish and aquatic plants. This body of research also encompasses animal feed regulations and requirements. It is important to regulate what feed is consumed by fish in order to prevent risks to human health and safety.

Seafood Safety Regulations

U.S. Labeling of Genetically Engineered Salmon

On November 19, 2015, the U.S. Food and Drug Administration (FDA) approved AquaBounty Technologies' application to sell the AquAdvantage salmon to U.S. consumers. The possible introduction of genetically engineered salmon into the marketplace furthers discussion involving ethics, protection of the natural environment, international and domestic trade law, labeling practices, nutrition, and constitutional issues. As the FDA points out, nutrition labeling for raw produce (fruits and vegetables), fish, and genetically modified products, is voluntary. Under section

403(a)(1) of the *Food, Drug, and Cosmetic Act* ("FD&C Act"), a food is misbranded if its labeling is "false or misleading in any particular". Section 201(n) of the FD&C Act provides that labeling is misleading if it fails to reveal facts that are material in light of representations made or suggested in the labeling. In regards to the AquAdvantage salmon, the FDA has stated:

> "Based on our assessments of food derived from the AquAdvantage Salmon, we have determined that the term "Atlantic salmon" is the appropriate common or usual name for such food within the meaning of section 403(i) of the FD&C Act because AquAdvantage Salmon meets FDA's regulatory standard for Atlantic salmon and the composition and basic nature of food from AquAdvantage Salmon does not significantly differ from its non-GE counterpart—non-GE farm-raised Atlantic salmon. In addition, we have determined that food derived from AquAdvantage Salmon is as safe and nutritious as food from other farm-raised Atlantic salmon. For these reasons, we have concluded that there is no material difference between food derived from AquAdvantage Salmon and food derived from other non-GE, farm-raised Atlantic salmon that is required to be disclosed in the labeling of food derived from AquAdvantage Salmon under the relevant provisions of the FD&C Act."

Canadian Labeling of Genetically Engineered Salmon

In 2014, Canada was a top-five producer of GM crops in the world, and is one of the largest players on the global market. Health Canada and the Canadian Food Inspection Agency ("CFIA") carry joint responsibility for federally food labeling policies in Canada under the *Food and Drugs Act* ("F&D Act"). Under the F&D Act, GMOs are defined as "novel foods". A novel food is allowed to enter the Canadian marketplace only after it has passed an assessment undertaken by various stakeholders. Health Canada is responsible for deciding all health and safety labeling policies for food products, such as special dietary needs or GM products, and ensuring that the products are safe for consumption. The CFIA is responsible for the development of all non-health and safety food labeling regulations and policies and the enforcement of labeling legislation. Standards are set by the CFIA so that Canadian food labels are truthful and not misleading.

Subsection 5(1) of the F&D Act states that "no person shall label, package, treat, process, sell or advertise any food in a manner that is false, misleading or deceptive or is likely to create an erroneous impression regarding its character, value, quantity, composition, merit or safety". Food must be labelled only if there are changes in the food such as problematic allergens or a significant nutrient or compositional change. Health Canada has not released an official statement concerning the AquAdvantage salmon, but does state on its website: "after twelve years of reviewing the safety of novel foods, Health Canada is not aware of any published scientific evidence demonstrating that novel foods are any less safe than traditional foods. The regulatory framework put in place by the federal government ensures that new and modified foods can be safely introduced in to the Canadian diet".

MARINE PROTECTED AREA

Marine Protected Area (MPA) is an umbrella term to describe a wide range of protected areas for marine conservation around the world. A global definition specifically for MPAs - as distinct from the general definition of a protected area - was first adopted by the IUCN in 1999. The definition was revised in 2012 and the distinction between a marine and terrestrial protected area was removed, aligning the definition of MPAs with the definition of a 'protected area' as "a clearly defined geographical space, recognised, dedicated and managed, through legal or other effective means, to achieve the long-term conservation of nature with associated ecosystem services and cultural values". To be included within the World Database on Protected Areas (WDPA), MPAs must be sites, located in the marine environment, that meet the most recent IUCN protected area definition.

The Convention on Biological Diversity (CBD) has a specific definition for marine and coastal protected areas (MCPAs) - "an area within or adjacent to the marine environment, together with its overlying waters and associated flora, fauna, and historical and cultural features, which has been reserved by legislation or other effective means, including custom, with the effect that its marine and/or coastal biodiversity enjoys a higher level of protection than its surroundings", which is aligned with the IUCN protected area definition, but is specifically designed to ensure that the advice given to CBD Parties is applicable to both marine and coastal areas.

All MPAs are designated for the purpose of conservation of biodiversity or cultural heritage. The designation of MPAs and MPA networks is driven by a range of international, regional, and national obligations and initiatives. Types of MPAs vary widely across regions but names for these MPA types (e.g. marine reserve, strict marine protected area) are not consistent between regions. MPA types can be used to describe the specific habitats they aim to protect (e.g. New Zealand's Benthic Protection Areas). More commonly, however, MPA types vary according to the protection being granted (e.g. no-access zones, no-take or no-impact zones). Although the legal designation of specific MPAs is done by national governments, communities may establish sites under their management, often termed Locally Managed Marine Areas (LMMAs), but these sites are not always officially recognised by their own national governments.

Although there is no single official instrument created to implement MPAs, the CBD is the most important international instrument addressing protected areas and the CBD Conference of Parties (CoP) agreed in 2004 that marine and coastal protected areas are an essential tool for delivering CBD targets on the conservation and sustainable use of marine and coastal biodiversity.

Coverage

There are approximately 15,000 MPAs, but since many of these are overlapping designations, the total coverage of MPAs is a better metric for measuring conservation progress. MPAs currently cover about 7% of the ocean's surface. This represents a quadrupling of overall MPA coverage over the last 10 years, but highlights the need for continued effort if CBD Aichi Target 11 is to be met, which aspires to protect at least 10% of the global marine and coastal environment within MPAs by 2020.

It is also worth noting that 20 of the largest MPAs comprise of 60% of the current MPA coverage. Furthermore, the vast majority of MPAs are in Territorial Seas (0-12 nautical miles from the coastline), whereas only just over 1% of Areas Beyond National Jurisdiction (beyond 200 nautical miles from the coastline) are covered by MPAs. While there has been rapid progress in creation of MPAs, there are concerns as to whether the Aichi target can be reached at the current rate of MPA establishment.

Criteria

Historically, MPAs were established on an ad hoc basis at the national level. There are no universally applicable criteria for establishing MPAs and MPA networks, however a number of national, regional and 'best practice' guidelines exist. Commonly mentioned criteria for the establishment of MPAs include naturalness, a high conservation value, feasibility, and a coherent management plan. Similarly, common criteria for establishing MPA networks include representativity of the wider area, connectivity of habitats and ecological processes and the replication of habitats/species in separate MPAs within the network.

Management

There are many different types of MPAs with the protection measures ranging from multiple-use to strict protection within 'no-take' or 'no-access' zones. Most MPAs tend to be more permissive, often implementing zones or seasonal restrictions that allow certain types of use.

The IUCN protected area management categories reflect the diverse range of purposes for which marine protected areas are declared. IUCN has published specific guidelines on how to apply its protected area management categories to MPAs, aiming to help authorities to maintain consistency in the ways that the management of both terrestrial and marine protected areas is described.

Business Relevance

Legal and compliance – Legal recognition and protection is likely to be present in these sites, although there are many areas important for marine conservation that are unprotected. Some MPAs are legally protected due to military interests. The protection of MPAs is also integrated into various policies and environmental safeguards of international financial institutions. These guidelines are regularly updated to incorporate recent developments in the field of conservation. IUCN has also published guidelines on the legislation of Protected Areas, which includes a section on the special considerations that must be taken into account when implementing legislation for MPAs. Such special considerations include the three-dimensionality of marine space, the exceptionally high connectivity between ecosystems, and the large proportion of ocean which falls in Areas Beyond National Jurisdiction (ABNJ).

Biodiversity importance – The MPAs are important areas of conservation of marine biodiversity and maintain productivity of oceans. The MPAs are site-scale units and are therefore highly relevant for mitigating and avoiding the risks of loss of marine biodiversity.

Socio-cultural values - Recognition of traditional practices, cultural values, rights and involvement of local/indigenous communities in protection, use and management is one of the criteria for identification of the MPAs. In establishment of several MPAs, local communities' rights to use and their role in management planning is emphasized.

FISH STOCKING

Fish Stocking is a fish management tool that works by releasing fish, usually bred in hatcheries, into the wild. Just like grocery items, it seems, the public has a demand for fish. Most often, the irate shoppers are replaced by angry anglers who want to be able to fish for sport. In other instances, environmentalists are the ones who want to restock empty waters to mitigate for past losses due to habitat disturbances or to restore historic populations. Sometimes, the customer is someone with a backyard pond, while at other times, the customer is a scientist who wants to conduct research.

The guidelines for stocking fish vary by location, with different countries and states having their own rules about which fish can be stocked where by whom. But whether it's a governmental office or a backyard enthusiast doing the stocking, careful planning is essential. After all, if a grocery store didn't plan its restocking schedule, the shelves might end up with no bread but hundreds of cans of peas. Imagine if two shoppers got into a brawl over that last loaf of bread, and you can start to imagine what might happen in an inappropriately planned fish stocking. While the shoppers may eventually settle the matter, the fish would die.

FISHING PROBLEMS

Management oversight, government regulations, traceability of fishing activities on the seas are a relatively recent development. But while some countries are now making a huge effort to stem overfishing, much more needs to be done.

A Host of Problems

In many cases, fisheries rules, regulations and enforcement measures are not efficient; fishing capacity and efforts are not sufficiently limited or controlled.

Another important issue is that today´s fishing activities often occur far from the eye of regulators and consumers.

Current key management problems include:

- Inadequate fisheries regulations: In many fisheries, current rules and regulations are not strong enough to limit fishing capacity to a sustainable level. This is particularly the case for the high seas, where there are few international fishing regulations.
- Lack of implementation/enforcement: Even when fisheries regulations exist, they are not always implemented or enforced. For example, many countries have still not ratified, implemented, or enforced international regulations such as the UN Convention on the Law of the Sea and the UN Fish Stocks Agreement. Lack of political will is also responsible for failures to adopt bycatch reduction devices.
- Lack of transparency and traceability: Customs agencies and also retailers are not always ensuring that the fish entering their country and shops is caught legally and in a sustainable way. As a result, consumers are unwittingly supporting poor management by purchasing fish from unsustainable fisheries. Only when our seafood is traceable can markets and legal systems be effective and reward sustainable practices, whilst deter the irresponsible.

- Failure to follow scientific advice: Many fisheries management bodies do not heed scientific advice on fish quotas and set catch limits above the recommended maximum amount; this is the case for for Atlantic cod and tuna, for example.
- Flag of Convenience vessels: Countries are either failing to restrict fishing companies from owning and operating FoC vessels, or are not rigorously inspecting FoC vessels landing at their ports. This include countries with some of the biggest fishing fleets such as the EU, Japan, Korea, and Taiwan (China). This allows illegal, unreported and unregulated (IUU) fishing to continue.
- Too few no-go areas for fishing: Protected areas and no-take zones, where fishing is banned or strictly regulated, can provide essential safe havens where young fish can grow to maturity and reproduce before they are caught. But just 1.2% of the world's oceans have been declared as Marine Protected Areas (MPAs), and 90% of existing MPAs are open to fishing. The current lack of protection is especially worrying for fish spawning grounds and the deep sea, both of which are particularly vulnerable to overfishing.

FISH DISEASE AND PARASITES

Bacterial Diseases of Fishes

Bacterial diseases are usually characterized by red streaks or spots and/or swelling of the abdomen or eye.

Red Pest

Symptoms:

Bloody streaks on body, fins and/or tail appear, so it is called red pest. In severe infection these streaks could lead to ulceration and possibly followed by fin and tail rot with the tail and/or fins falling off.

Treatment:

External treatments are usually not effective as the disease is internal.

At the Appearance of Disease

- Treat the tank with a disinfectant and clean the tank as best as possible.
- To disinfect, use acriflavine (trypaflavine) or monacrin (mono-amino-acridine) using a 0.2% solution at the rate of 1 ml per litre. Both disinfectants will colour the water, but the colour disappears as the disinfectants dissipate.
- Do not feed a lot while the fish is being treated.
- If the fish do not appear to respond favorably, discontinue disinfecting. Then add an

antibiotic to the food. 1% of antibiotic may be carefully mixed with flake food. If you keep the fish hungry they should eagerly eat the mixture before the antibiotic dissipates.

Antibiotics usually available in 250 mg capsules. If added to 25 grams of flake food, one capsule should be enough to treat dozens of fishes. A good antibiotic are Chloromycetin (chloramphenicol) and use tetracycline.

Mouth Fungus

Symptoms:

White cottony patches around the mouth. It looks like a fungus attack of the mouth, so it is called mouth fungus. It is actually caused from the bacterium Chondrococcus columnaris. In the beginning a grey or white line appear around the lips and later short tufts arise from the mouth like fungus. This disease may be fatal due to production of toxins and the inability to eat. Hence treatment at an early stage is essential.

Treatment:

Penicillin at 10000 units per litre is a very effective treatment. Second dose should be given in two days, or use Chloromycetin, 10 to 20 mg per litre, with a second dose in two days.

Tuberculosis

Symptoms:

Emaciation, hollow belly, possibly sores. Tuberculosis is caused by the Bacterium Mycobacterium piscium. Fish infected with tuberculosis may become hollow bellied, pale, show skin ulcers and frayed fins, and loss of appetite. Yellowish or darker nodules may appear on the body or eyes. The main cause for this disease appears to be over-crowding in un-kept conditions.

Treatment:

There is no known and effective treatment for this disease. The best thing to do is to destroy the infected fish and, if un-kept conditions or overcrowding is the suspected cause, it is required to take necessary measures.

Dropsy

Symptoms:

Bloating of the body, protruding scales. Dropsy is caused from a bacterial infection (acromonas) of the kidneys, causing fluid accumulation or renal failure. The fluids in the body build up and cause the fish to bloat up and the scales to protrude.

Treatment:

An effective treatment is to add an antibiotic to the food. With flake food, use about 1% of antibiotic and carefully mix it. Antibiotics in 250 mg capsules if added to 25 grams of flake food will be sufficient to treat dozens of fishes. A good antibiotic is Chloromycetin (chloramphenicol), or use tetracycline.

Scale Protrusion

Symptoms:

Protruding scales without body bloat. Bacterial infection of the scales and/or body causes scale protrusion. An effective treatment is to add an antibiotic to the food. With flake food, use about 1% of antibiotic like Chloromycetin (chloramphenicol), or tetracycline.

Tail Rot and Fin Rot

Symptoms:

Disintegrating fins that may be reduced to stumps, exposed fin rays, blood on edges of fins, reddened areas at base of fins, skin ulcers with grey or red margins, cloudy eyes. It is caused by the bacteria Aeromonas. If tank conditions are not good an infection can be caused from a simple injury to the fins/tail. Tuberculosis can lead to tail and fin rot. Basically, the tail and/or fins become frayed or lose colour.

Treatment:

Treat the water or fish with antibiotics. A good antibiotic is Chloromycetin (chloramphenicol) or tetracycline. Treatment of 1% $CuSO_4$ is also effective.

Ulcer

Symptoms:

Loss of appetite and slow body movements. It is caused by baceria, haemophilus.

Treatment:

Give dip treatment in 1% $CUSO_4$ for one minute for a period of 3 to 4 days. Antibiotics oxytetracycline and chloramphenicol may be useful in acute infection.

Precautions during Treatment of Bacterial Diseases

Bacterial diseases are best treated by antibiotics such as penicillin, amoxicillin, or erythromycin. The most common parasitic disease called "Ich" can be treated most effectively with copper or malachite green in the right dosage.

Most medications contain copper as an ingredient. Many water treatments like "Aquari-Sol" will also contain copper as an ingredient. The copper may be harmful to most plants and invertebrates, such as snails. Indeed, most snail removers are copper based.

Antibiotic may disturb biological filtration in the tank. Therefore, it is also recommended to monitor either ammonia and nitrite levels of water, or use an ammonia remover such as "Am-Quel" to be sure that the level of ammonia does not exceeds the desired limit.

Fungal Diseases of Fishes

Argulosis

Symptoms:

Caused due to Argulus (Fish louse). The fish scrapes itself against objects, clamped fins, visible parasites about 1/4 inch in diameter are visible on the body of the fish. The fish louse is a flattened mite-like crustacean about 5 mm long that attaches itself to the body of fish. They irritate the host fish which may have clamped fins, become restless, and may show inflamed areas where the lice have been.

Treatment:

With larger fish and light infestations, the lice can be picked off with a pair of forceps. Other cases can best be done with a 10 to 30 minute bath in 10 mg per litre of potassium permanganate or treat the whole tank with 2 mg per liter, but this method is messy and dyes the water.

Ichthyosporidium

Symptoms:

Sluggishness, loss of balance, hollow belly, external cysts and sores.

Ichthyosporidium is a fungus, but it manifests itself internally. It primarily attacks the liver and kidneys, but it spread everywhere else. The symptoms vary. The fish may become sluggish, loosely balanced, show hollow bellies, and eventually show external cysts or sores. By then it is usually too late for the fish.

Treatment:

Phenoxethol added to food as a 1% solution may be effective. Chloromycetin added to the food has also been effective. But both of these treatments, if not watched with caution, could pose a risk to your fish. It is best, if diagnosed soon enough, to destroy the affected fish before the disease spreads.

Fungus (Saprolegnia)

Symptoms:

Tufts of dirty, cotton-like growth on the skin, can cover large areas of the fish, fish eggs turn white. Fungal attacks always follow some other health problems like parasitic attack, injury, or bacterial infection. The symptoms are a grey or whitish growth in and on the skin and/or fins of the fish.

Eventually, if left untreated, these growths will become cottony looking. The fungus, if left untreated, will eventually eat away on the fish until it finally dies.

Treatment:

Use a solution of phenoxethol at 1% in distilled water. Add 10 ml of this solution per litre. Repeat after a few days if needed, but only once more as three treatments could be dangerous inhabitants.

If the symptoms are severe the fish can be removed and treated with small amount of providone iodine or mercurochrome. For attacks on fish eggs, most breeders will use a solution of methylene blue adding 3 to 5 mg/1 as a preventive measure after the eggs are laid.

Parasitic Diseases of Fishes

Velvet or Rust

Symptoms:

Yellow to light brown "dust" on body, clamped fins, respiratory distress (breathing hard). This disease has the appearance of a golden or brownish dust over the fins and body. The fish may show signs of irritation, like glancing off aquarium decor, shortage of breath (fish-wise), and clamping of the fins.

The gills are usually the first thing affected. Velvet affects different species in different ways. Danios seem to be the most susceptible, but often show no discomfort. The disease is highly contagious and fatal.

Treatment:

The best treatment is with copper at 0.2 mg per litre (0.2 ppm) to be repeated once in a few days, if necessary. Acriflaving (trypaflavine) may be used instead at 0.2% solution (1 ml per liter). As acriflavine can possibly sterilize fish and copper can lead to poisoning, the water should be gradually changed after a cure has been effected.

Anchor Worm (Lernaea)

Symptoms:

The fish scrapes itself against objects, whitish-green threads hang out of the fish's skin with an inflamed area at the point of attachment. Ahchor worms are actually crustaceans.

The youngs are free swimming and borrow into the skin, go into the muscles and develop for several months before showing. They release eggs and die. The holes left behind are ugly and may become infected. The anchor worm is too deeply imbedded to safely remove.

Treatment:

A 10 to 30 minute bath in 10 mg per litre of potassium permanganate, or treat the whole tank with 2 mg per litre, but this method is messy and dyes the water.

Ergasilus

Symptoms:

The fish scrapes itself against objects, whitish-green threads hang out of the fish's gills. This parasite is like the anchor worm, but is smaller and attacks the gills instead of the skin.

Treatment:

Treatment can best be done with a 10 to 30 minute bath in 10 mg per litre of potassium permanganate.

Flukes

Symptoms:

The fish scrapes itself against objects, rapid till movement, mucus covering the gills or body, the gills or fins may be eaten away, the skin may become reddened. There are many species of flukes, which are flatworms about 1 mm long, and several symptoms that are visible. They infest gills and skin much like Ich, but the difference can be seen with a hand lens.

You should be able to see movement and possibly eyespots, which is not found in Ich. Gill flukes will eventually destroy the gills thus killing the fish. Symptoms of heavy infestations are pale fish with drooping fins, rapid respiration, glancing off aquarium decor, and/or hollow bellies.

Treatment:

Treatment can best be done with a 10 to 30 minute bath in 10 mg per litre of potassium permanganate. Or treat the whole tank with 2 mg per litre, but this method is messy and dyes the water.

Nematoda

Symptoms:

Worms hanging from the anus. Nematodes (threadworms) infect just about anywhere in the body but only shows itself when they hang out of the anus. A heavy infestation causes hollow bellies.

Treatment:

Two treatments have been suggested. First treatment: soak the food in para-chloro-meta-xylenol and give the fish a bath or treat the aquarium with 10 ml per litre. The bath should last for several days. Second treatment: find special food containing thiabendazole as a nematode (threadworm) cure and hope the fish will eat it.

Leeches

Symptoms:

Leeches are visible on the fish's skin. Leeches are external parasites and affix themselves on the body, fins, or gills of the fish. Usually they appear as heart shaped worms (they are just curled up) attached to the fish. Since leeches are sucking and borrowing into the surface of the fish, removal with forceps can cause great damage, if not death, to the fish.

Treatment:

If the fish is bathed in a 2.5 percent solution of salt for 15 minutes, most of the leeches should just fall off. Those that do not will be affected enough to remove with forceps with minimal damage. Another treatment is to add Trichlorofon at 0.25 mg/l.

Protozoan Diseases in Fishes

Costia

Symptoms:

Milky cloudiness on skin.

Treatment:

This is a rare protozoan disease that causes a cloudiness of the skin. The best treatment is with copper at 0.2 mg per liter (0.2 ppm) to be repeated once in a few days, if necessary. Acriflavine (trypaflavine) may be used instead at 0.2% solution (1 ml per litre). As acriflavine can possibly sterilize fish, and copper can lead to poisoning, the water should be gradually changed after a cure has been effected.

Hexamita

Symptoms:

Caused by intestinal flagellated protozoa that attack the lower intestine. As it is a disease of the digestive tract, characterized by the loss of appetite.

Treatment:

An effective treatment is the drug metronidazole. A combined treatment in the food (1% in any food the fish will eat) and in the water (12 mg per liter) is recommended. Repeat the water treatment every other day for three treatments.

Ich (Ichthyphthirious)

Symptoms:

Salt-like specks on the body fins. Excessive slime. Problems in breathing (Ich invades the gills), clamped fins, loss of appetite. Ich, white spot disease, whatever the name, this is the most common malady experienced in the home aquarium. Luckily, this disease is also easily cured if noticed in time. Ich is actually a protozoa called Ichthyophthirious multifiliis.

There are three phases to the life cycle of these protozoa. Normally, to the amateur aquarist, the life cycle is of no importance. However, since Ich is susceptible to treatment at only one stage of the life cycle, an awareness of the life cycle is important.

Adult Phase

It is embedded in the skin or gills of the fish, causing irritation (with the fish showing signs of irritation) and the appearance of small white nodules. As the parasite grows it feeds on red blood cells and skin cells. After a few days it bores itself out of the fish and falls to the bottom of the aquarium.

Cyst Phase

After falling to the bottom, the adult parasite forms into a cyst with rapid cell divisions occurring.

Free Swimming Phase

After the cyst phase, about 1000 free swimming youngs swim upwards looking for a host. If a host is not found within 2 to 3 days, the parasite dies. Once a host is found the whole cycle begins a-new.

Treatment:

The drug of choice is quinine hydrochloride at 30 mg per litre (1 in 30000). Quinine sulphate can be used if the hydrochloride is not available. The water may cloud but this will disappear. By reducing the time (with raised temperature) of the phases, you should be able to attack the free swimming phase effectively. Most commercial remedies contain malachite green and /or copper, which are both effective.

Neon Tetra Disease

Symptoms:

Whitened areas deep into the fishes flesh. Muscle degeneration leading to abnormal swimming movements. So named for the fish it was first recognized on. It is caused by the sporozoa Plistophora hyphessobryconis. Even though it is named after Neon Tetras, it can appear on other fish. Whitish patches appear as if just below the skin.

In Neon Tetras it destroys the bright blue-green neon stripe. The organisms form cysts which burst and release spores. The spores penetrate further and form more cysts. Eventually, the spores migrate to the water and are eaten by other fish in the food. These spores migrate into the digestive tract, then the muscles, and a new infection starts.

Treatment:

There is no known cure. It is best to destroy the infected fish and clean the aquarium.

Glugea and Henneguya

Symptoms:

Similar to Lymphocytes, the fish will have nodular white swellings on fins or body.

Treatment:

Glugea and Hnneguya and sporozoans that form large cysts on the fish's body and release spores. Luckily, these diseases are very rare. The fish bloat up, with tumour- like protrusions, and eventually die. No cure, as of yet. It is best to destroy the infected fish before the spores spread.

Chilodonella

Symptoms:

Dulling of the colours due to excessive slime, frying of the fins, weakness, gill damage. This disease causes a blue white cloudiness on the skin and attacks the gills. Later the skin may be broken down and the gills destroyed. The fish may behave as if they have irritations.

Treatment:

Acriflaving (trypaflaving) may be used at 1% solution (5 ml per litre). As acriflavine can sterilize fish, the water should be gradually changed after a cure has been effected. It also helps to raise the temperature to about 80 °F.

Whirling Disease

Symptoms:

It is also a protozoan disease, caused by Myxosoma cerebralis. Blakeening of tail, caudal band and deformity of anal region are common symptoms.

Treatment:

Destroy all the diseased fishes by applying quicklime at the rate of 1kg/ha.

Knot Disease

Symptoms:

It is caused by protozoa, Myxobolus exigus. Salt konts appear on the skin.

Treatment:

There is no effective treatment. Therefore, all infected fishes should be removed immediately and killed.

Bio-disease

Symptom:

It is caused by protozoa, Myxobolus pfcifferi. Large boils of varying size of nut appear in several parts of the body.

Treatment:

Give bath in 3% common salt solution or in 1% formalin solution for 10 minutes.

Myxosporidisis

Symptoms:

It is caused by infection of Myxosorida. Cysts appear on the body, internal tissues and organs. Fish becomes weak. Scales become weak, perforated and fall off.

Treatment:

Give dip treatment in 10% common salt solution.

Non-infectious Maladies in Fishes

Tumours

Tumours can be caused by a virus or a cancer, but most tumours are genetic. The genetic tumours may be caused from too much hybridization, common amongst professional breeders. Practically all tumours are untreatable. If the fish is in distress, it should be destroyed.

Congenital Abnormalities

Abnormalities usually occur when professional breeders are trying to acquire certain strains in breeds.

Physical Injuries

As in the human world, accidents happen in the lives of fishes also. If the cause of the injury is obvious, it should be remedied. Then the injury should be treated. Then injury should be touched with 2% mercurochrome, which is supplied commercially.

Also, depending on the fish's tolerance to water conditions, keeping the fish in slightly acid- water should speed recovery (pH 6.6). Minor injuries, if the water conditions are good, should just heal themselves.

Constipation

Some fishes are more susceptible to constipation than others. Usually fish with more compressed bodies like angelfish and silver dollars. Symptoms are loss of appetite and swelling of the body. The cause is almost always diet.

Usually, with a change of diet, the condition rights itself. But in stubborn cases try dried food that has been soaked in medicinal paraffin oil. Glycerol or castor oil may also be used. If the diet is changed on a regular basis and live foods offered are occasionally, this condition may never occur.

The intensification and further development of freshwater aquaculture in India urgently requires knowledge, research facilities and research and expertise on fish diseases and fish health protection.

Miscellaneous Diseases in Fishes

Head and Lateral Line Disease (also known as Hole-in-the-Head disease):

Symptoms:

Begins as small pits on the head and face, usually just above the eye. If untreated, these turn into large cavities and then the disease progresses along the lateral line. Head and Lateral Line Disease is attributed to a nutritional deficiency of one or more of vitamin C, vitamin D, calcium, and phosphorous.

It is thought to be caused by a poor diet or lack of variety, lack of partial water changes or over filtration with chemical media such as activated carbon.

Treatment:

- Increase frequent water changes.
- Add vitamins to frozen foods.

- Add flake foods, as they are enriched with vitamins.
- Add green, either frozen or in leaf form, to the diet.
- Decrease the amount of beef heart as it lacks many critical nutrients.

This disease is often confused with another disease called Hexamita, because both these diseases are often seen simultaneously in the same fish. Haxamita is a protozoan disease that attacks the lower intestine.

Eye Problems

Symptoms:

Cloudy cornea, opaque lens, poor eye, swelling, blindness.

- Cloudy cornea can result from a bacterial invasion. Antibiotics may help.
- Opaqueness can result from poor nutrition or a metacercaria invasion (grubs). Try foods with added vitamins and change the diet to include variety.
- Pop eye (exophthalmia) can result from rough handling, gas embolism, tumours, bacterial infection, or vitamin A deficiency. Gas bubble or bacterial infection can be treated successfully with penicillin or amoxicillin.
- Blindness can be caused by poor nutrition or excessive light. Lowering the light level and a change of diet to include lots of variety may help prevent it.

Swim-bladder Disease

Symptoms:

Abnormal swimming pattern, difficulty in maintaining equilibrium.

If you suspect swim-bladder problems in a fish, first check and treat it for other diseases as listed below:

- Congenitally deformed bladder.
- Cancer or tuberculosis in organs adjacent to swim bladder.
- Constipation.
- Poor nutrition.
- Serious parasitic and bacterial infestation.

Viral Disease in Fishes

Lymphocytis

Symptoms:

Nodular white swellings (cauliflower) on fins or body. Lymphocystis is a virus and being a virus, affects the cells of the fish. It usually manifests itself as abnormally large white lumps (cauliflower) on the fins or other parts of the body. It can be infectious but is usually not fatal. Unfortunately, there is no cure. Fortunately, this is a rare disease.

Treatment:

There are two suggested treatments. One treatment is to remove and destroy the infected fish as soon as possible. The other treatment is to simply separate the infected fish foe several months and wait for remission, which usually does occur.

References

- Fisheries-management, entry: newworldencyclopedia.org, Retrieved 16 January, 2019
- Fisheries-management-techniques, commercial, fishing: gov.au, Retrieved 29 March, 2019
- Understanding-ecosystem-based-fisheries-management, insight: noaa.gov, Retrieved 30 April, 2019
- Marine-protected-area-mpa: biodiversitya-z.org, Retrieved 29 June, 2019
- Fish-stocking, fish-populations, fish-conservation, outdoor-activities: howstuffworks.com, Retrieved 14 July, 2019
- Fisheries-management, problems, oceans, our-work: panda.org, Retrieved 11 January, 2019
- 7-major-diseases-of-fish-with-prevention-and-treatment-fish-farming, applied-fisheries, fish: yourarticlelibrary.com, Retrieved 18 April, 2019

Threats to Fisheries

CHAPTER 6

Sustainable fishery refers to the harvesting of fishes at a sustainable rate such that fish population does not decline over time. Some of the challenges to sustainable fishery are overfishing, illegal fishing and marine pollution. The topics elaborated in this chapter will help in gaining a better perspective about sustainable fishery and its challenges.

Sustainability means ensuring the habitat and ecosystem supporting the fishery are in good condition. A sustainable fishery has sufficient spawning fish to produce the next generation, while allowing fishing to take place. Sustainable fishing means leaving enough fish in the ocean, respecting habitats and ensuring people who depend on fishing can maintain their livelihoods.

Sustainable fishing guarantees there will be populations of ocean and freshwater wildlife for the future. Aquatic environments are home to countless species of fish and invertebrates, most of which are consumed as food. (Others are harvested for economic reasons, such as oysters that produce pearls used in jewelry.) Seafood is respected all over the world, in many diverse cultures, as an important source of protein and healthy fats. For thousands of years, people have fished to feed families and local communities.

Demand for seafood and advances in technology have led to fishing practices that are depleting fish and shellfish populations around the world. Fishers remove more than 77 billion kilograms (170 billion pounds) of wildlife from the sea each year. Scientists fear that continuing to fish at this rate may soon result in a collapse of the world's fisheries. In order to continue relying on the ocean as an important food source, economists and conservationists say we will need to employ sustainable fishing practices.

Consider the example of the bluefin tuna. This fish is one of the largest and fastest on Earth. It is known for its delicious meat, which is often enjoyed raw, as sushi. Demand for this particular fish has resulted in very high prices at markets and has threatened its population. Today's spawning population of bluefin tuna is estimated at 21 to 29 percent of its population in 1970.

Since about that time, commercial fishers have caught bluefin tuna using purse seining and longlining. Purse seine fishing uses a net to herd fish together and then envelop them by pulling the net's drawstring. The net can scoop up many fish at a time, and is typically used to catch schooling fish or those that come together to spawn. Longlining is a type of fishing in which a very long line—up to 100 kilometers (62 miles)—is set and dragged behind a boat. These lines have thousands of baited hooks attached to smaller lines stretching downward.

Both purse seining and longlining are efficient fishing methods. These techniques can catch hundreds or thousands of fish at a time.

Sustainable Fishing Practices

There are ways to fish sustainably, allowing us to enjoy seafood while ensuring that populations remain for the future. In many indigenous cultures, people have fished sustainably for thousands of years. Today's sustainable fishing practices reflect some lessons learned from these cultures.

In the Philippines, the Tagbanua people have traditionally employed fishing practices that simultaneously harvest and maintain fish populations. They continue to follow these practices today. Tagbanuas fish for specific species only during certain times of the year, determined by tides and the moon, allowing fish stocks to replenish themselves. They set aside certain areas, such as coral reefs, as protected spots in which fishing is prohibited. When they do fish, these traditional fishers primarily use hook-and-line methods, catching only what they need to feed themselves and their communities. A 2007 study lauded traditional Tagbanua practices as a way to prevent injury and death to local Irrawaddy dolphins, which become entangled in more modern fishing gear like nets and traps.

Traditional Polynesian cultures of the South Pacific have also always relied on the ocean's resources. Their most common historical fishing practices were hook and line, spearfishing, and cast nets. Hooks constructed of bone, shell, or stone were designed to catch specific species. Fishers would also craft 2-meter (6-foot) spears. They would dive underwater or spear fish from above, again targeting specific animals. Cast nets were used by fishers working individually or in groups. The nets could be cast from shore or canoes, catching groups of fish. All of these methods targeted fish needed for fishers' families and local communities.

Some of these sustainable fishing practices are still used today. Native Hawaiians practice cast-net fishing and spearfishing. Modern spearfishing is practiced all over the world, including in South America, Africa, Australia, and Asia. In many cases, spear guns are now used to propel the spear underwater. Spearfishing is a popular recreational activity in some areas of the United States, including Florida and Hawaii. This fishing method is considered sustainable because it targets one fish at a time and results in very little by catch.

If you have ever gone fishing, chances are you used a rod and reel. Rod-and-reel fishing is a modern version of traditional hook-and-line. Rods and reels come in different shapes and sizes, allowing recreational and commercial fishers to target a wide variety of fish species in both freshwater and saltwater. The different types of rods and reels, coupled with different locations and bait, mean fishers can catch pelagic fish like sailfish, bottom-dwellers like flounder, and freshwater species such as catfish and trout. Rod-and-reel fishing results in less bycatch because non-targeted species can be released immediately. Additionally, only one fish is caught at a time, preventing overfishing. For commercial fishers, rod-and reel-fishing is a more sustainable alternative to long lining.

Another way to prevent overfishing and bycatch is to simply abstain from eating fish and other seafood. Dr. Sylvia Earle, renowned marine scientist and National Geographic Explorer-in-Residence, suggests people need to take a break from eating seafood until we learn better how to maintain healthy fish and wildlife populations.

> "I personally have stopped eating seafood," she explained to National Geographic. "I know too much. I know that every fish counts at this point. Some more than others, but I can no longer bear the thought of eating tuna knowing in what dire straits they currently are. If we

value the ocean and the ocean's health at all, we have to understand that fish are critical to maintaining the integrity of ocean systems, which in turn make the planet work."

Reconciling Fisheries with Conservation

Management goals might consider the impact of salmon on bear and river ecosystems.

At the Fourth World Fisheries Congress in 2004, Daniel Pauly asked, "How can fisheries science and conservation biology achieve a reconciliation?", then answered his own question, "By accepting each other's essentials: that fishing should remain a viable occupation; and that aquatic ecosystems and their biodiversity are allowed to persist."

A relatively new concept is relationship farming. This is a way of operating farms so they restore the food chain in their area. Re-establishing a healthy food chain can result in the farm automatically filtering out impurities from feed water and air, feeding its own food chain, and additionally producing high net yields for harvesting. An example is the large cattle ranch Veta La Palma in southern Spain. Relationship farming was first made popular by Joel Salatin who created a 220 hectare relationship farm featured prominently in Michael Pollan's book *The Omnivore's Dilemma* (2006) and the documentary films, Food, Inc. and Fresh. The basic concept of relationship farming is to put effort into building a healthy food chain, and then the food chain does the hard work.

Overfishing

Overfishing can be sustainable. According to Hilborn, overfishing can be "a misallocation of societies' resources", but it does not necessarily threaten conservation or sustainability".

Overfishing is traditionally defined as harvesting so many fish that the yield is less than it would be if fishing were reduced. For example, Pacific salmon are usually managed by trying to determine how many spawning salmon, called the "escapement", are needed each generation to produce the maximum harvestable surplus. The optimum escapement is that needed to reach that surplus. If the escapement is half the optimum, then normal fishing looks like overfishing. But this is still sustainable fishing, which could continue indefinitely at its reduced stock numbers and yield. There is a wide range of escapement sizes that present no threat that the stock might collapse or that the stock structure might erode.

On the other hand, overfishing can precede severe stock depletion and fishery collapse. Hilborn points out that continuing to exert fishing pressure while production decreases, stock collapses and the fishery fails, is largely "the product of institutional failure."

Today over 70% of fish species are either fully exploited, overexploited, depleted, or recovering from depletion. If overfishing does not decrease, it is predicted that stocks of all species currently commercially fished for will collapse by 2048."

A Hubbert linearization (Hubbert curve) has been applied to the whaling industry, as well as charting the price of caviar, which depends on sturgeon stocks. Another example is North Sea cod. Comparing fisheries and mineral extraction tells us that human pressure on the environment is causing a wide range of resources to go through a Hubbert depletion cycle.

Coastal fishing communities in Bangladesh are vulnerable to flooding from sea-level rises.

Island with fringing reef in the Maldives. Coral reefs are dying around the world.

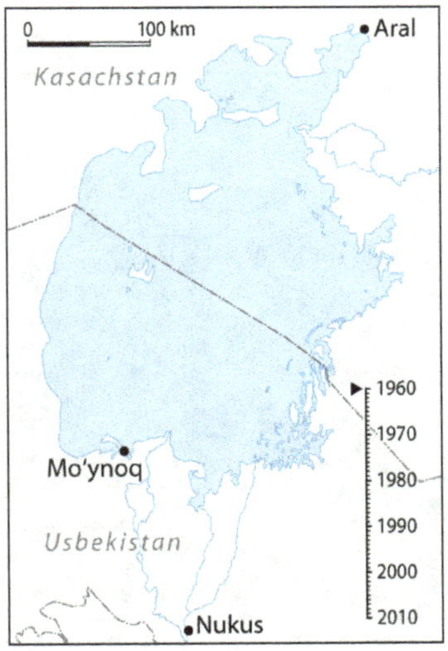

Shrinking of the Aral Sea.

Habitat Modification

Nearly all the world's continental shelves, and large areas of continental slopes, underwater ridges, and seamounts, have had heavy bottom trawls and dredges repeatedly dragged over their surfaces. For fifty years, governments and organizations, such as the Asian Development Bank, have encouraged the fishing industry to develop trawler fleets. Repeated bottom trawling and dredging literally flattens diversity in the benthic habitat, radically changing the associated communities.

Climate Change

Rising ocean temperatures and ocean acidification are radically altering aquatic ecosystems. Climate change is modifying fish distribution and the productivity of marine and freshwater species. This reduces sustainable catch levels across many habitats, puts pressure on resources needed for aquaculture, on the communities that depend on fisheries, and on the oceans' ability to capture and store carbon (biological pump). Sea level rise puts coastal fishing communities at risk, while changing rainfall patterns and water use impact on inland (freshwater) fisheries and aquaculture.

Ocean Pollution

A recent survey of global ocean health concluded that all parts of the ocean have been affected by human development and that 41 percent has been fouled with human polluted runoff, overfishing, and other abuses. Pollution is not easy to fix, because pollution sources are so dispersed, and are built into the economic systems we depend on.

The United Nations Environment Programme (UNEP) mapped the impacts of stressors such as climate change, pollution, exotic species, and over-exploitation of resources on the oceans. The report shows at least 75 percent of the world's key fishing grounds may be affected.

Diseases and Toxins

Large predator fish can contain significant amounts of mercury, a neurotoxin which can affect fetal development, memory, mental focus, and produce tremors.

Irrigation

Abandoned ship near Aral, Kazakhstan.

Lakes are dependent on the inflow of water from its drainage basin. In some areas, aggressive irrigation has caused this inflow to decrease significantly, causing water depletion and a shrinking of the lake. The most notable example is the Aral Sea, formerly among the four largest lakes in the world, now only a tenth of its former surface area.

Remediation

Fisheries Management

Fisheries management draws on fisheries science to enable sustainable exploitation. Modern fisheries management is often defined as mandatory rules based on concrete objectives and a mix of management techniques, enforced by a monitoring control and surveillance system.

- Ideas and rules: Economist Paul Romer believes sustainable growth is possible providing the right ideas (technology) are combined with the right rules, rather than simply hectoring fishers. There has been no lack of innovative ideas about how to harvest fish. He characterizes failures as primarily failures to apply appropriate rules.
- Fishing subsidies: Government subsidies influence many of the world fisheries. Operating cost subsidies allow European and Asian fishing fleets to fish in distant waters, such as West Africa. Many experts reject fishing subsidies and advocate restructuring incentives globally to help struggling fisheries recover.
- Economics: Another focus of conservationists is on curtailing detrimental human activities by improving fisheries' market structure with techniques such as salable fishing quotas, like those set up by the Northwest Atlantic Fisheries Organization.
- Valorization of by-catch: Helping to avoid discards (and their associated adverse ecological impacts) by valorizing by-catch products, as they are good sources for protein hydrolizates, peptones, enzymatic mixtures or fish oil being these products of interest different industrial sectors.
- Payment for Ecosystem Services: Environmental Economist, Essam Y Mohammed, argues that by creating direct economic incentives, whereby people are able to receive payment for

the services their property provides, will help to establish sustainable fisheries around the world as well as inspire conservation where it otherwise would not.

- Sustainable fisheries certification: A promising direction is the independent certification programs for sustainable fisheries conducted by organizations such as the Marine Stewardship Council and Friend of the Sea. These programs work at raising consumer awareness and insight into the nature of their seafood purchases.

Ecosystem-based Fisheries

According to marine ecologist Chris Frid, the fishing industry points to marine pollution and global warming as the causes of recent, unprecedented declines in fish populations. Frid counters that overfishing has also altered the way the ecosystem works. "Everybody would like to see the rebuilding of fish stocks and this can only be achieved if we understand all of the influences, human and natural, on fish dynamics." He adds: "fish communities can be altered in a number of ways, for example they can decrease if particular-sized individuals of a species are targeted, as this affects predator and prey dynamics. Fishing, however, is not the sole cause of changes to marine life—pollution is another example. No one factor operates in isolation and components of the ecosystem respond differently to each individual factor."

The traditional approach to fisheries science and management has been to focus on a single species. This can be contrasted with the ecosystem-based approach. Ecosystem-based fishery concepts have been implemented in some regions. In a 2007 effort to "stimulate much needed discussion" and "clarify the essential components" of ecosystem-based fisheries science, a group of scientists offered the following ten commandments for ecosystem-based fisheries scientists:

- Keep a perspective that is holistic, risk-adverse and adaptive.
- Maintain an "old growth" structure in fish populations, since big, old and fat female fish have been shown to be the best spawners, but are also susceptible to overfishing.
- Characterize and maintain the natural spatial structure of fish stocks, so that management boundaries match natural boundaries in the sea.
- Monitor and maintain seafloor habitats to make sure fish have food and shelter.
- Maintain resilient ecosystems that are able to withstand occasional shocks.
- Identify and maintain critical food-web connections, including predators and forage species.
- Adapt to ecosystem changes through time, both short-term and on longer cycles of decades or centuries, including global climate change.
- Account for evolutionary changes caused by fishing, which tends to remove large, older fish.
- Include the actions of humans and their social and economic systems in all ecological equations.

Marine Protected Areas

Strategies and techniques for marine conservation tend to combine theoretical disciplines, such as population biology, with practical conservation strategies, such as setting up protected areas, as with Marine Protected Areas (MPAs) or Voluntary Marine Conservation Areas. Each nation defines MPAs independently, but they commonly involve increased protection for the area from fishing and other threats.

Marine life is not evenly distributed in the oceans. Most of the really valuable ecosystems are in relatively shallow coastal waters, above or near the continental shelf, where the sunlit waters are often nutrient rich from land runoff or upwellings at the continental edge, allowing photosynthesis, which energizes the lowest trophic levels. In the 1970s, for reasons more to do with oil drilling than with fishing, the U.S. extended its jurisdiction, then 12 miles from the coast, to 200 miles. This made huge shelf areas part of its territory. Other nations followed, extending national control to what became known as the exclusive economic zone (EEZ). This move has had many implications for fisheries conservation, since it means that most of the most productive maritime ecosystems are now under national jurisdictions, opening possibilities for protecting these ecosystems by passing appropriate laws.

Daniel Pauly characterises marine protected areas as "a conservation tool of revolutionary importance that is being incorporated into the fisheries mainstream." The Pew Charitable Trusts have funded various initiatives aimed at encouraging the development of MPAs and other ocean conservation measures.

Fish Farming

There exists concerns that farmed fish cannot produce necessary yields efficiently. For example, farmed salmon eat three pounds of wild fish to produce one pound of salmon.

Laws and Treaties

International laws and treaties related to marine conservation include the 1966 Convention on Fishing and Conservation of Living Resources of the High Seas. United States laws related to marine conservation include the 1972 Marine Mammal Protection Act, as well as the 1972 Marine Protection, Research and Sanctuaries Act which established the National Marine Sanctuaries program. Magnuson-Stevens Fishery Conservation and Management Act.

Awareness Campaigns

Introducing the results of long term monitoring to a local fishermen in Kihnu, Estonia.

Various organizations promote sustainable fishing strategies, educate the public and stakeholders, and lobby for conservation law and policy. The list includes the Marine Conservation Biology Institute and Blue Frontier Campaign in the U.S. The U.K.'s Frontier (the Society for Environmental Exploration) and Marine Conservation Society, Australian Marine Conservation Society, International Council for the Exploration of the Sea (ICES), Langkawi Declaration, Oceana, PROFISH, and the Sea Around Us Project, International Collective in Support of Fishworkers, World Forum of Fish Harvesters and Fish Workers, Frozen at Sea Fillets Association and CEDO.

The United Nations Millennium Development Goals include, as goal 7: target 2, the intention to "reduce biodiversity loss, achieving, by 2010, a significant reduction in the rate of loss", including improving fisheries management to reduce depletion of fish stocks.

Some organizations certify fishing industry players for sustainable or good practices, such as the Marine Stewardship Council and Friend of the Sea.

Other organizations offer advice to members of the public who eat with an eye to sustainability. According to the marine conservation biologist Callum Roberts, four criteria apply when choosing seafood:

- Is the species in trouble in the wild where the animals were caught?
- Does fishing for the species damage ocean habitats?
- Is there a large amount of bycatch taken with the target species?
- Does the fishery have a problem with discards—generally, undersized animals caught and thrown away because their market value is low?

Data Issues

Data Quality

One of the major impediments to the rational control of marine resources is inadequate data. According to fisheries scientist Milo Adkison, the primary limitation in fisheries management decisions is poor data. Fisheries management decisions are often based on population models, but the models need quality data to be accurate. Scientists and fishery managers would be better served with simpler models and improved data.

Unreported Fishing

Estimates of illegal catch losses range between $10 billion and $23 billion annually, representing between 11 and 26 million tonnes.

- Incidental catch

Shifting Baselines

Shifting baselines is the way significant changes to a system are measured against previous baselines, which themselves may represent significant changes from the original state of the system.

The term was first used by the fisheries scientist Daniel Pauly in his paper "Anecdotes and the shifting baseline syndrome of fisheries". Pauly developed the term in reference to fisheries management where fisheries scientists sometimes fail to identify the correct "baseline" population size (e.g. how abundant a fish species population was *before* human exploitation) and thus work with a shifted baseline. He describes the way that radically depleted fisheries were evaluated by experts who used the state of the fishery at the start of their careers as the baseline, rather than the fishery in its untouched state. Areas that swarmed with a particular species hundreds of years ago, may have experienced long term decline, but it is the level of decades previously that is considered the appropriate reference point for current populations. In this way large declines in ecosystems or species over long periods of time were, and are, masked. There is a loss of perception of change that occurs when each generation redefines what is "natural".

Looting the Seas

Looting the seas is the name given by the International Consortium of Investigative Journalists to a series of journalistic investigations into areas directly affecting the sustainability of fisheries. So far they have investigated three areas involving fraud, negligence and overfishing:

- The black market in bluefin tuna.
- Subsidies propping up the Spanish fishing industry.
- Overfishing of the southern jack mackerel.

Other Factors

The focus of sustainable fishing is often on the fish. Other factors are sometimes included in the broader question of sustainability. The use of non-renewable resources is not fully sustainable. This might include diesel fuel for the fishing ships and boats: there is even a debate about the long term sustainability of biofuels. Modern fishing nets are usually made of artificial polyamides like nylon. Synthetic braided ropes are generally made from nylon, polyester, polypropylene or high performance fibers such as high modulus polyethylene (HMPE) and aramid.

Energy and resources are employed in fish processing, refrigeration, packaging, logistics, etc. The methodologies of Life-cycle assessment are useful to evaluate the sustainability of components and systems. These are part of the broad question of sustainability.

OVERFISHING

Overfishing is the removal of a species of fish from a body of water at a rate that the species cannot replenish in time, resulting in those species either becoming depleted or very underpopulated in that given area. Overfishing has spread all over the globe and has been present for centuries.

Overfishing can occur in water bodies of any sizes, such as ponds, rivers, lakes or oceans, and can result in resource depletion, reduced biological growth rates and low biomass levels. Sustained overfishing can lead to critical depensation, where the fish population is no longer able to sustain itself. Some forms of overfishing, such as the overfishing of sharks, has led to the upset of entire marine ecosystems.

The ability of a fishery to recover from overfishing depends on whether the ecosystem's conditions are suitable for the recovery. Dramatic changes in species composition can result in an ecosystem shift, where other equilibrium energy flows involve species compositions different from those that had been present before the depletion of the original fish stock. For example, once trout have been overfished, carp might take over in a way that makes it impossible for the trout to re-establish a breeding population.

Global Scale

Overfishing has stripped many fisheries around the world of their stocks. The United Nations Food and Agriculture Organization estimated in a 2018 report that 33.1% of world fish stocks are subject to overfishing. Significant overfishing has been observed in pre-industrial times. In particular, the overfishing of the western Atlantic Ocean from the earliest days of European colonisation of the Americas has been well documented.

The fraction of fish stocks that are within biologically sustainable levels has exhibited a decreasing trend, from 90.0 percent in 1974 to 66.9 percent in 2015. In contrast, the percentage of stocks fished at biologically unsustainable levels increased from 10 percent in 1974 to 33.1 percent in 2015, with the largest increases in the late 1970s and 1980s.

In 2015, maximally sustainably fished stocks (formerly termed fully fished stocks) accounted for 59.9 percent and underfished stocks for 7.0 percent of the total assessed stocks. While the proportion of underfished stocks decreased continuously from 1974 to 2015, the maximally sustainably fished stocks decreased from 1974 to 1989, and then increased to 59.9 percent in 2015.

In 2015, among the 16 major statistical areas, the Mediterranean and Black Sea had the highest percentage (62.2 percent) of unsustainable stocks, closely followed by the Southeast Pacific 61.5 percent and Southwest Atlantic 58.8 percent. In contrast, the Eastern Central Pacific, Northeast Pacific (Area 67), Northwest Pacific (Area 61), Western Central Pacific and Southwest Pacific had the lowest proportion (13 to 17 percent) of fish stocks at biologically unsustainable levels.

Daniel Pauly, a fisheries scientist known for pioneering work on the human impacts on global fisheries, has commented:

> "It is almost as though we use our military to fight the animals in the ocean. We are gradually winning this war to exterminate them. And to see this destruction happen, for nothing really – for no reason – that is a bit frustrating. Strangely enough, these effects are all reversible, all the animals that have disappeared would reappear, all the animals that were small would grow, all the relationships that you can't see any more would re-establish themselves, and the system would re-emerge."

Examples: Examples of overfishing exist in areas such as the North Sea, the Grand Banks of Newfoundland and the East China Sea. In these locations, overfishing has not only proved disastrous to fish stocks but also to the fishing communities relying on the harvest. Like other extractive industries such as forestry and hunting, fisheries are susceptible to economic interaction between ownership or stewardship and sustainability, otherwise known as the tragedy of the commons.

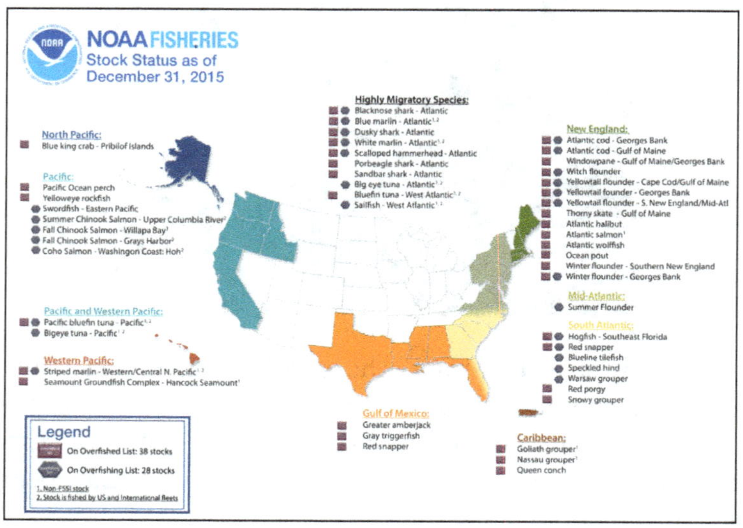

Overfished US stocks, 2015.

- The Peruvian coastal anchovy fisheries crashed in the 1970s after overfishing and an El Niño season largely depleted anchovies from its waters. Anchovies were a major natural resource in Peru; indeed, 1971 alone yielded 10.2 million metric tons of anchovies. However, the following five years saw the Peruvian fleet's catch amount to only about 4 million tons. This was a major loss to Peru's economy.

- The collapse of the cod fishery off Newfoundland, and the 1992 decision by Canada to impose an indefinite moratorium on the Grand Banks, is a dramatic example of the consequences of overfishing.
- The sole fisheries in the Irish Sea, the west English Channel, and other locations have become overfished to the point of virtual collapse, according to the UK government's official Biodiversity Action Plan. The United Kingdom has created elements within this plan to attempt to restore this fishery, but the expanding global human population and the expanding demand for fish has reached a point where demand for food threatens the stability of these fisheries, if not the species' survival.
- Many deep sea fish are at risk, such as orange roughy and sablefish. The deep sea is almost completely dark, near freezing and has little food. Deep sea fish grow slowly because of limited food, have slow metabolisms, low reproductive rates, and many do not reach breeding maturity for 30 to 40 years. A fillet of orange roughy at the store is probably at least 50 years old. Most deep sea fish are in international waters, where there are no legal protections. Most of these fish are caught by deep trawlers near seamounts, where they congregate because of food. Flash freezing allows the trawlers to work for days at a time, and modern fishfinders target the fish with ease.
- Blue walleye became extinct in the Great Lakes in the 1980s. Until the middle of the 20th century, it was a commercially valuable fish, with about a half million tonnes being landed during the period from about 1880 to the late 1950s, when the populations collapsed, apparently through a combination of overfishing, anthropogenic eutrophication, and competition with the introduced rainbow smelt.
- The World Wide Fund for Nature and the Zoological Society of London jointly issued their "Living Blue Planet Report" on 16 September 2015 which states that there was a dramatic fall of 74% in worldwide stocks of the important scombridae fish such as mackerel, tuna and bonitos between 1970 and 2010, and the global overall "population sizes of mammals, birds, reptiles, amphibians and fish fell by half on average in just 40 years."

In Management

Several countries are now effectively managing their fisheries. Examples include Iceland and New Zealand. The United States has turned many of its fisheries around from being in a highly depleted state.

Consequences

According to a 2008 UN report, the world's fishing fleets are losing US$50 billion each year through depleted stocks and poor fisheries management. The report, produced jointly by the World Bank and the UN Food and Agriculture Organization (FAO), asserts that half the world's fishing fleet could be scrapped with no change in catch. In addition, the biomass of global fish stocks have been allowed to run down to the point where it is no longer possible to catch the amount of fish that could be caught. Increased incidence of schistosomiasis in Africa has been linked to declines of fish species that eat the snails carrying the disease-causing parasites. Massive growth of jellyfish populations threaten fish stocks, as they compete with fish for food, eat fish eggs, and poison or swarm fish, and can survive in oxygen depleted environments where fish cannot; they wreak massive havoc on commercial fisheries. Overfishing eliminates a major jellyfish competitor and predator exacerbating the jellyfish population explosion. Both climate change and a restructuring of the ecosystem have been found to be major roles in an increase in jellyfish population in the Irish Sea in the 1990s.

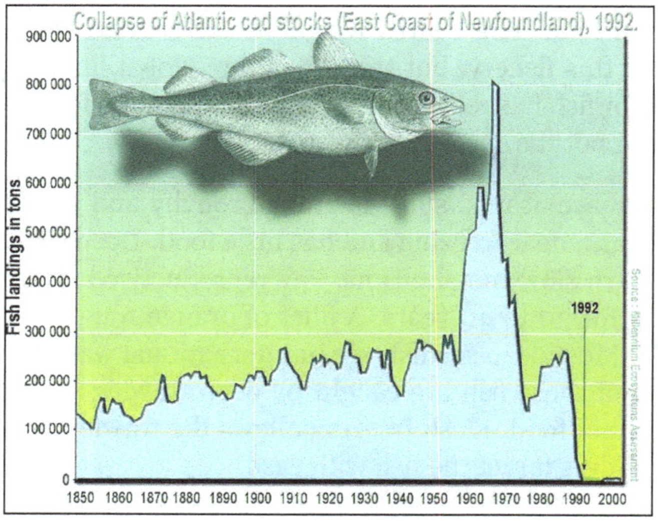

Atlantic cod stocks were severely overfished in the 1970s and 1980s, leading to their abrupt collapse in 1992.

Types

There are three recognized types of biological overfishing: growth overfishing, recruit overfishing and ecosystem overfishing.

Growth

Growth overfishing occurs when fish are harvested at an average size that is smaller than the size that would produce the maximum yield per recruit. A recruit is an individual that makes it to maturity, or into the limits specified by a fishery, which are usually size or age. This makes the total yield less than it would be if the fish were allowed to grow to an appropriate size. It can be countered by reducing fishing mortality to lower levels and increasing the average size of harvested fish to a size that will allow maximum yield per recruit.

Recruitment

Recruitment overfishing occurs when the mature adult population (spawning biomass) is depleted to a level where it no longer has the reproductive capacity to replenish itself—there are not enough adults to produce offspring. Increasing the spawning stock biomass to a target level is the approach taken by managers to restore an overfished population to sustainable levels. This is generally accomplished by placing moratoriums, quotas and minimum size limits on a fish population.

Ecosystem

Ecosystem overfishing occurs when the balance of the ecosystem is altered by overfishing. With declines in the abundance of large predatory species, the abundance of small forage type increases causing a shift in the balance of the ecosystem towards smaller fish species.

Acceptable Levels

The notion of overfishing hinges on what is meant by an acceptable level of fishing. More precise biological and bioeconomic terms define acceptable level as follows:

- Biological overfishing occurs when fishing mortality has reached a level where the stock biomass has negative marginal growth (reduced rate of biomass growth), as indicated by the red area in the figure. (Fish are being taken out of the water so quickly that the replenishment of stock by breeding slows down. If the replenishment continues to diminish for long enough, replenishment will go into reverse and the population will decrease.)
- Economic or bioeconomic overfishing additionally considers the cost of fishing when determining acceptable catches. Under this framework, a fishery is considered to be overfished when catches exceed maximum economic yield where resource rent is at its maximum. Fish are being removed from the fishery so quickly that the profitability of the fishery is sub-optimal. A more dynamic definition of economic overfishing also considers the present value of the fishery using a relevant discount rate to maximise the flow of resource rent over all future catches.

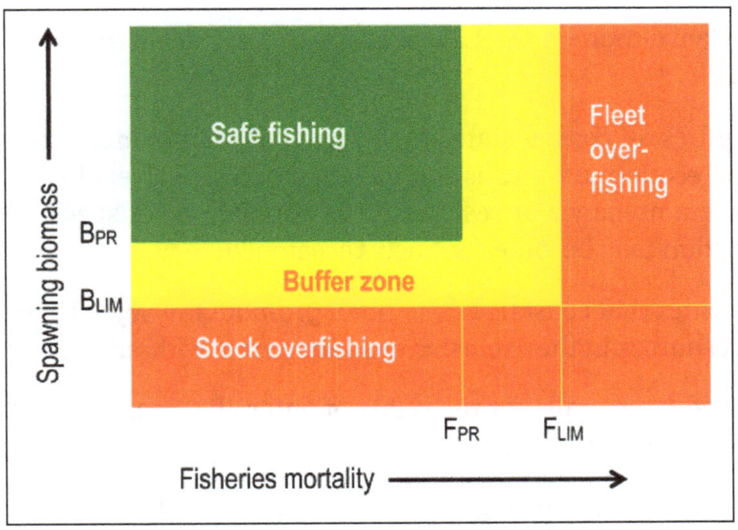

The Traffic Light colour convention, showing the concept of Harvest Control Rule (HCR), specifying when a rebuilding plan is mandatory in terms of precautionary and limit reference points for spawning biomass and fishing mortality rate.

Harvest Control Rule

A model proposed in 2010 for predicting acceptable levels of fishing is the Harvest Control Rule (HCR), which is a set of tools and protocols with which management has some direct control of harvest rates and strategies in relation to predicting stock status, and long-term maximum sustainable yields. Constant catch and constant fishing mortality are two types of simple harvest control rules.

Input and Output Orientations

Fishing capacity can also be defined using an input or output orientation.

- An input-oriented fishing capacity is defined as the maximum available capital stock in a fishery that is fully utilized at the maximum technical efficiency in a given time period, given resource and market conditions.
- An output-oriented fishing capacity is defined as the maximum catch a vessel (fleet) can produce if inputs are fully utilized given the biomass, the fixed inputs, the age structure of the fish stock, and the present stage of technology.

Technical efficiency of each vessel of the fleet is assumed necessary to attain this maximum catch. The degree of capacity utilization results from the comparison of the actual level of output (input) and the capacity output (input) of a vessel or a fleet.

Mitigation

In order to meet the problems of overfishing, a precautionary approach and Harvest Control Rule (HCR) management principles have been introduced in the main fisheries around the world. The Traffic Light color convention introduces sets of rules based on predefined critical values, which could be adjusted as more information is gained.

The United Nations Convention on the Law of the Sea treaty deals with aspects of over fishing in articles 61, 62, and 65.

- Article 61 requires all coastal states to ensure that the maintenance of living resources in their exclusive economic zones is not endangered by over-exploitation. The same article addresses the maintenance or restoration of populations of species above levels at which their reproduction may become seriously threatened.
- Article 62 provides that coastal states: "shall promote the objective of optimum utilization of the living resources in the exclusive economic zone without prejudice to Article 61".
- Article 65 provides generally for the rights of, inter alia, coastal states to prohibit, limit, or regulate the exploitation of marine mammals.

According to some observers, overfishing can be viewed as an example of the tragedy of the commons; appropriate solutions would therefore promote property rights through, for instance, privatization and fish farming. Daniel K. Benjamin, in *Fisheries are Classic Example of the "Tragedy of the Commons"*, cites research by Grafton, Squires and Fox to support the idea that privatization can solve the overfishing problem:

> "According to recent research on the British Columbia halibut fishery, where the commons has been at least partly privatized, substantial ecological and economic benefits have resulted. There is less damage to fish stocks, the fishing is safer, and fewer resources are needed to achieve a given harvest."

Another possible solution, at least for some areas, is quotas, so fishermen can only legally take a certain amount of fish. A more radical possibility is declaring certain areas of the sea "no-go zones" and make fishing there strictly illegal, so the fish in that area have time to recover and repopulate.

In order to maximise resources some countries (e.g. Bangladesh and Thailand) have improved the availability of family planning services. The resulting smaller populations have a decreased environmental footprint and less food needs to provided.

Controlling consumer behavior and demand is critical in mitigating action. Worldwide, a number of initiatives emerged to provide consumers with information regarding the conservation status of the seafood available to them. The Guide to Good Fish Guides lists a number of these.

Government Regulation

Many regulatory measures are available for controlling overfishing. These measures include fishing quotas, bag limits, licensing, closed seasons, size limits and the creation of marine reserves and other marine protected areas.

A model of the interaction between fish and fishers showed that when an area is closed to fishers, but there are no catch regulations such as individual transferable quotas, fish catches are temporarily increased but overall fish biomass is reduced, resulting in the opposite outcome from the one desired for fisheries. Thus, a displacement of the fleet from one locality to another will generally have little effect if the same quota is taken. As a result, management measures such as temporary closures or establishing a marine protected area of fishing areas are ineffective when not combined with individual fishing quotas. An inherent problem with quotas is that fish populations vary from year to year. A study has found that fish populations rise dramatically after stormy years due to more nutrients reaching the surface and therefore greater primary production. To fish sustainably, quotas need to be changed each year to account for fish population.

Individual transferable quotas (ITQs) are fishery rationalization instruments defined under the Magnuson-Stevens Fishery Conservation and Management Act as limited access permits to harvest quantities of fish. Fisheries scientists decide the optimal amount of fish (total allowable catch) to be harvested in a certain fishery. The decision considers carrying capacity, regeneration rates and future values. Under ITQs, members of a fishery are granted rights to a percentage of the total allowable catch that can be harvested each year. These quotas can be fished, bought, sold, or leased allowing for the least cost vessels to be used. ITQs are used in New Zealand, Australia, Iceland, Canada, and the United States.

In 2008, a large-scale study of fisheries that used ITQs compared to ones that didn't provided strong evidence that ITQs can help to prevent collapses and restore fisheries that appear to be in decline.

China bans fishing in the South China Sea for a period each year.

Removal of Subsidies

Several scientists have called for an end to subsidies paid to deep sea fisheries. In international waters beyond the 200 nautical mile exclusive economic zones of coastal countries, many fisheries are unregulated, and fishing fleets plunder the depths with state-of-the-art technology. In a few hours, massive nets weighing up to 15 tons, dragged along the bottom by deep-water trawlers, can destroy deep-sea corals and sponge beds that have taken centuries or millennia to grow. The trawlers can target orange roughy, grenadiers, or sharks. These fish are usually long-lived and late maturing, and their populations take decades, even centuries to recover.

Fisheries scientist Daniel Pauly and economist Ussif Rashid Sumaila have examined subsidies paid to bottom trawl fleets around the world. They found that US$152 million per year are paid to deep-sea fisheries. Without these subsidies, global deep-sea fisheries would operate at a loss of $50 million a year. A great deal of the subsidies paid to deep-sea trawlers is to subsidize the large amount of fuel required to travel beyond the 200-mile limit and drag weighted nets.

> "There is surely a better way for governments to spend money than by paying subsidies to a fleet that burns 1.1 billion litres of fuel annually to maintain paltry catches of old growth fish from highly vulnerable stocks, while destroying their habitat in the process" – *Pauly*.
>
> "Eliminating global subsidies would render these fleets economically unviable and would relieve tremendous pressure on over-fishing and vulnerable deep-sea ecosystems" – *Sumaila*.

Minimizing Fishing Impact

Fishing techniques may be altered to minimize bycatch and reduce impacts on marine habitats. These techniques include using varied gear types depending on target species and habitat type. For example, a net with larger holes will allow undersized fish to avoid capture. A turtle excluder device (TED) allows sea turtles and other megafauna to escape from shrimp trawls. Avoiding fishing in spawning grounds may allow fish stocks to rebuild by giving adults a chance to reproduce.

Aquaculture

Aquaculture involves the farming of fish in captivity. This approach effectively privatizes fish stocks and creates incentives for farmers to conserve their stocks. It also reduces environmental impact. However, farming carnivorous fish, such as salmon, does not always reduce pressure on wild fisheries, since carnivorous farmed fish are usually fed fishmeal and fish oil extracted from wild forage fish.

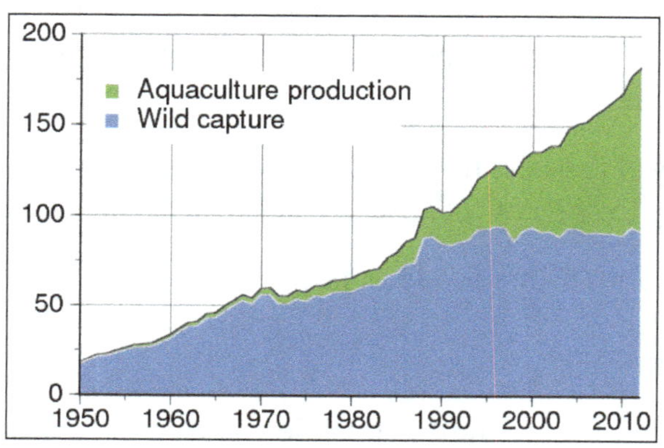

Global harvest of aquatic organisms in million tonnes, 1950–2010, as reported by the FAO.

Aquaculture played a minor role in the harvesting of marine organisms until the 1970s. Growth in aquaculture increased rapidly in 1990s when the rate of wild capture plateaued. Aquaculture now provides approximately half of all harvested aquatic organisms. Aquaculture production rates continue to grow while wild harvest remains steady.

Fish farming can enclose the entire breeding cycle of the fish, with fish being bred in captivity. Some fish prove difficult to breed in captivity and can be caught in the wild as juveniles and brought into captivity to increase their weight. With scientific progress more species are being made to breed in captivity. This was the case with southern bluefin tuna, which were first bred in captivity in 2009.

Consumer Awareness

Sustainable seafood is a movement that has gained momentum as more people become aware of overfishing and environmentally destructive fishing methods. Sustainable seafood is seafood from either fished or farmed sources that can maintain or increase production in the future without jeopardizing the ecosystems from which it was acquired. In general, slow-growing fish that reproduce late in life, such as orange roughy, are vulnerable to overfishing. Seafood species that grow quickly and breed young, such as anchovies and sardines, are much more resistant to overfishing. Several organizations, including the Marine Stewardship Council (MSC), and Friend of the Sea, certify seafood fisheries as sustainable.

The Marine Stewardship Council has developed an environmental standard for sustainable and well-managed fisheries. Environmentally responsible fisheries management and practices are rewarded with the use of its blue product ecolabel. Consumers concerned about overfishing and its consequences are increasingly able to choose seafood products that have been independently assessed against the MSC's environmental standard. This enables consumers to play a part in reversing the decline of fish stocks. As of February 2012, over 100 fisheries around the world have been independently assessed and certified as meeting the MSC standard. Their where-to-buy page lists the currently available certified seafood. As of February 2012 over 13,000 MSC-labelled products are available in 74 countries around the world. Fish & Kids is an MSC project to teach schoolchildren about marine environmental issues, including overfishing.

The Monterey Bay Aquarium's Seafood Watch Program, although not an official certifying body like the MSC, also provides guidance on the sustainability of certain fish species. Some seafood restaurants have begun to offer more sustainable seafood options. The Seafood Choices Alliance is an organization whose members include chefs that serve sustainable seafood at their establishments. In the US, the Sustainable Fisheries Act defines sustainable practices through national standards. Although there is no official certifying body like the MSC, the National Oceanic and Atmospheric Administration has created FishWatch to help guide concerned consumers to sustainable seafood choices.

In September 2016, a partnership of Google and Oceana and Skytruth introduced Global Fishing Watch, a website designed to assist citizens of the globe in monitoring fishing activities.

Barriers to Effective Management

The fishing industry has a strong financial incentive to oppose some measures aimed at improving the sustainability of fish stocks. Recreational fisherman also have an interest in maintaining access to fish stocks. This leads to extensive lobbying that can block or water down government policies intended to prevent overfishing.

Outside of countries' exclusive economic zones, fishing is difficult to control. Large oceangoing fishing boats are free to exploit fish stocks at will.

In waters that are the subject of territorial disputes, countries may actively encourage overfishing. A notable example is the cod wars where Britain used its navy to protect its trawlers fishing in Iceland's exclusive economic zone. Fish are highly transitory. Many species will freely move through different jurisdictions. The conservation efforts of one country can then be exploited by another.

While governments can create regulations to control people's behaviors this can be undermined by illegal fishing activity. Estimates of the size of the illegal catch range from 11 to 26 million tonnes, which represents 14-33% of the world's reported catch. Illegal fishing can take many forms. In some developing countries, large numbers of poor people are dependent on fishing. It can prove difficult to regulate this kind of overfishing, especially for weak governments. Even in regulated environments, illegal fishing may occur. While industrial fishing is often effectively controlled, smaller scale and recreational fishermen can often break regulations such as bag limits and seasonal closures. Fisherman can also easily fish illegally by doing things such as underreporting the amount of fish they caught or reporting that they caught one type of fish while actually catching another. There is also a large problem with surveillance of illegal fishing activity. In 2001, the UN Food and Agriculture Organization (FAO), passed the International Plan of Action to Prevent, Deter and Eliminate Illegal, Unreported and Unregulated Fishing (IPOA-IUU). This is an agreement with the intention to stop port states from allowing boats to dock that participated in illegal, unreported or unregulated fishing. It also gives details for port states on effective measures of inspecting and reporting illegal fishing. Some illegal fishing takes place on an industrial scale with financed commercial operations.

The fishing capacity problem is not only related to the conservation of fish stocks but also to the sustainability of fishing activity. Causes of the fishing problem can be found in the property rights regime of fishing resources. Overexploitation and rent dissipation of fishermen arise in open-access fisheries .

In open-access resources like fish stocks, in the absence of a system like individual transferable quotas, the impossibility of excluding others provokes the fishermen who want to increase catch to do so effectively by taking someone else' share, intensifying competition. This tragedy of the commons provokes a capitalization process that leads them to increase their costs until they are equal to their revenue, dissipating their rent completely.

FISHING DOWN THE FOOD WEB

Fishing down the food web is the process whereby fisheries in a given ecosystem, "having depleted the large predatory fish on top of the food web, turn to increasingly smaller species, finally ending up with previously spurned small fish and invertebrates".

The process was first demonstrated by the fisheries scientist Daniel Pauly and others in an article published in the journal *Science* in 1998. Large predator fish with higher trophic levels have been depleted in wild fisheries. As a result, the fishing industry has been systematically "fishing down the food web", targeting fish species at progressively decreasing trophic levels.

The trophic level of a fish is the position it occupies on the food chain. In 2000, the Convention on Biological Diversity selected the mean trophic level of fisheries catch, renamed the "Marine Trophic Index" (MTI), as one of eight indicators of ecosystem health. However, many of the world's most lucrative fisheries are crustacean and mollusk fisheries, which are at low trophic levels and thus result in lower MTI values.

The mean trophic level of the world fisheries catch has steadily declined because many high trophic level fish, such as this tuna, have been overfished.

Fishermen are increasingly targeting lower trophic level fish, like these anchovies and other forage fish.

According to Daniel Pauly, if the trend continues, consumers may be eating jellyfish sandwiches.

Over the last 50 years, the abundance of large predator fish, such as cod, swordfish and tuna, has dropped 90 percent. Fishing vessels now increasingly pursue the smaller forage fish, such as herrings, sardines, menhaden and anchovies, that are lower on the food chain. "We are eating bait and moving on to jellyfish and plankton" says Pauly. Beyond this, the overall global volume of fish captured has been declining since the late 1980s.

Mean Trophic Level

The mean trophic level is calculated by assigning each fish or invertebrate species a number based on its trophic level. The trophic level is a measure of the position of an organism in a food web, starting at level 1 with primary producers, such as phytoplankton and seaweed, then moving through the primary consumers at level 2 that eat the primary producers to the secondary consumers at level 3 that eat the primary consumers, and so on. In marine environments, trophic levels range from two to five for the apex predators. The mean trophic level can then be calculated for fishery catches by averaging trophic levels for the overall catch using the datasets for commercial fish landings.

Ecopath

Pauly's team used the catch data from the FAO which it fed into an Ecopath model. Ecopath is a computerised ecosystem modelling system. The functioning of an ecosystem can be described using path analysis to track the direction and influence of the many factors controlling the ecosystem. The original Ecopath model was applied to a coral reef food web. Scientists tracked tiger sharks at the top of the food web and collected data on their feeding behavior, what they ate and how much. Likewise, they collected feeding data on the other organisms in the food chains down to the primary producers, such as algae. This data was fed into an Ecopath model, which then described the energy flow, in terms of food, as it moved from the primary producers up the food web to the apex predator. Such models allow scientists to compute the complex effects that occur, both direct and indirect, from the interactions of the many ecosystem components.

The model showed that over the last 50 years the mean trophic level of fish catches has declined by somewhere between 0.5 and 1.0 trophic levels. This decline applied both globally, on a worldwide scale, and more locally on a scale specific to oceans, that is, for the separate FAO subareas: the Atlantic, Indian and Pacific Oceans, and the Mediterranean-Black Seas.

Marine Trophic Index

In 2000, the Convention on Biological Diversity, an international treaty aimed at sustaining biodiversity which has been adopted by 193 member countries, selected the mean trophic level of fisheries catch as one of eight indicators for immediate testing. They renamed it the "Marine Trophic Index" (MTI), and have mandated that member countries report over time on changes in ocean trophic levels as a primary indicator of marine biodiversity and health.

The Marine Trophic Index is a measure of the overall health and stability of a marine ecosystem or area. The index is also a proxy measure for overfishing and an indication of how abundant and rich the large, high trophic level fish are.

Changes in the Marine Trophic Index over time can function as an indicator of the sustainability of a country's fish resources. It can indicate the extent that the fishing effort within a country's fishing grounds is modifying its fish stocks. A negative change generally indicates that larger predator fish are becoming depleted, and an increasing number of smaller forage fish are being caught. A zero or positive change in the Marine Trophic Index indicates the fishery is stable or improving.

Ecology of Fishing Down

Ecologically, the decline in the mean trophic level is explained by the relationship between the size of the fish captured and their trophic level. The trophic level of fishes usually increases with their size, and fishing tends to selectively capture the larger fishes. This applies both between species as well as within species. When the fishing is intense, the relative abundance of the larger fish positioned high in the food chain is reduced. Consequently, over time, small fishes start to dominate the fisheries catches, and the mean trophic level of the catches declines. Recently the market value of small forage fishes and invertebrates, which have low trophic levels, has sharply increased to the point where they can be considered to be subsidizing fishing down.

Daniel Pauly has suggested a framework for the ecological impacts fishing down can have on marine ecosystems. The framework distinguishes three phases:

- Pristine – the first phase. Pristine environments are the states ocean ecosystem were in before fishing made strong impacts. Some outlier areas of the South Pacific may still be pristine. For most of the world what these pristine states may have been can only be inferred from archaeological data, historical accounts and anecdotes. In pristine ocean environments, the biomass of the large predator fish are 10 to 100 times greater than their present biomass. This implies a large supporting biomass of small prey fishes and invertebrates. On the seafloor, the benthos is dominated by deposit feeders which prevent resuspension of sediments, and filter feeders which keep the phytoplankton down. Thus the water column tends to be oligotrophic, free of both suspended particles and of the nutrients that leach from them.

- Exploited – the second phase. Exploited is the phase we are currently in. It is characterized by declines in the biomass of large predator fish, declines in the diversity, size and trophic level of captured fish, and declines in the benthos. Bottom trawlers progressively destroy the biogenic structures built over many years on the seafloor by the filter and detritus feeders. As these structures and animals that were filtering the phytoplankton and consuming the detritus (marine snow) disappear, they are replaced with the polyp stages of jellyfish and other small errant benthic animals. Storms resuspend the marine snow, and the water column gradually eutrophies. In the early part of this phase, cascade effects compensate these declines with the emergence of new fisheries for opportunistic feeders, such as squid, shrimp and other invertebrates. But eventually this decline also.

- Fully degraded – the third phase. The dead zone is the biological endpoint of a fully degraded marine ecosystem. The dead zone is a zone with excessive nutrients in the water column, resulting in the depletion of oxygen and the elimination of multicellular organisms. The abundant detritus and marine snow are processed by bacteria rather than by the benthic animals. These dead zones are currently growing throughout the world in places such as the Bohai Sea in China, the northern Adriatic Sea, and the northern Gulf of Mexico. Some estuaries, such as the Chesapeake Bay estuary, also display features associated with a fully degraded marine ecosystem. In the Chesapeake Bay, overfishing eliminated the benthic filter feeders such as oysters, and most predators larger than a striped bass, the current apex predator. One hundred and fifty years ago, the oysters formed giant reefs and filtered the waters of Chesapeake Bay every three days. Because the oysters are gone, pollution entering the estuary from rivers now produces harmful algal blooms.

DESTRUCTIVE FISHING PRACTICES

Destructive fishing practices are practices that easily result in irreversible damage to aquatic habitats and ecosystems. Many fishing techniques can be destructive if used inappropriately, but some practices are particularly likely to result in irreversible damage. These practices are mostly, though not always, illegal. Where they are illegal, they are often inadequately enforced.

The narrowest definition of destructive fishing practices refers principally to bottom trawling over vulnerable habitat (shallow corals, deep sea corals, or seagrass, for example), as well as practices such as shark finning, blast fishing, poison fishing, muroami, and push netting. These latter practices are not significant within the fishing zones of most developed nations, being generally outlawed.

A wider and more useful definition would include:

- Overfishing beyond reasonable recovery limits, including serial overfishing.
- Excessive or damaging levels of bycatch.
- The fishing of spawning aggregations without precautionary justification.
- Intensive fishing over vulnerable habitats, including for example spawning and nursery areas.
- Highgrading–the practice of discarding the lower quality portion of the target catch.

This definition could be extended to cover activities such as:

- Ghost fishing by lost or discarded gear.
- Shark netting of popular swimming beaches (with high incidental catch).
- Amateur use of fish aggregating devices or traps where they increase the likelihood of locally unsustainable catch levels.
- Spearfishing at night or with SCUBA, as well as 'industrial' spearfishing.
- Use of stainless steel hooks or traps.
- Commercial and recreational use of gill nets and traps with high incidental kill.

- Deliberate (and sometimes illegal) destruction of marine life perceived as "getting in the way" of fishing operations.

The phrase destructive fishing practices (or DFPs) has been featured in international fisheries literature since about the 1980s. No widely accepted definition of the phrase exists, and this will almost certainly remain the situation, given very different national and industry perspectives. The DFPs discussed with the most standing are those of the Fisheries and Agriculture Organization of the United Nations - the FAO.

The Outcomes and Implementation Statements of the *World Summit on Sustainable Development*, held in Johannesburg in 2002, contain a commitment to phasing out destructive fishing practices in the marine environment by the year 2012. All nations attending the summit supported this statement.

Many nations had made commitments to end destructive fishing practices much earlier. In 1999, 124 nations explicitly gave their support to the FAO *Code of Conduct for Responsible Fisheries 1995* through the Rome Declaration on Responsible Fisheries. The list of these nations includes most of the major fishing nations of the world. However, while the Code of Conduct contains a commitment to end destructive fishing practices, the Code contains no timelines.

Blast Fishing

Dynamite or blast fishing is done easily and cheaply with dynamite or homemade bombs made from locally available materials. Fish are killed by the shock from the blast and are then skimmed from the surface or collected from the bottom. The explosions indiscriminately kill large numbers of fish and other marine organisms in the vicinity and can damage or destroy the physical environment. Explosions are particularly harmful to coral reefs. Blast fishing is also illegal in many waterways around the world.

Bottom Trawling

Bottom trawling is trawling (towing a trawl, which is a fishing net) along the sea floor. It is also referred to as "dragging". The scientific community divides bottom trawling into benthic trawling and demersal trawling. Benthic trawling is towing a net at the very bottom of the ocean and demersal trawling is towing a net just above the benthic zone. Bottom trawling targets both bottom-living fish (groundfish) and semi-pelagic species such as cod, squid, shrimp, and rockfish.

Bottom fishing has operated for over a century on heavily fished grounds such as the North Sea and Grand Banks. While overfishing has long been recognised as causing major ecological changes to the fish community on the Grand Banks, concern has been raised more recently about the damage which benthic trawling inflicts upon seabed communities. A species of particular concern is the slow growing, deep water coral *Lophelia pertusa*. This species is home to a diverse community of deep sea organisms, but is easily damaged by fishing gear. On 17 November 2004, the United Nations General Assembly urged nations to consider temporary bans on high seas bottom trawling.

Bottom trawling over vulnerable habitat, however, will continue within the Exclusive Fishing Zones of most nations until governments have mapped the location of vulnerable habitats, and taken steps to exclude all bottom trawling activities from these areas.

Cyanide Fishing

Cyanide fishing is a method of collecting live fish mainly for use in aquariums, which involves spraying a sodium cyanide mixture into the desired fish's habitat in order to stun the fish. The practice hurts not only the target population, but also many other marine organisms, including coral and thus coral reefs.

Recent studies have shown that the combination of cyanide use and stress of post capture handling results in mortality of up to 75% of the organisms within less than 48 hours of capture. With such high mortality numbers, a greater number of fish must be caught in order to offset post catch death.

Muroami

Muroami is a destructive artisan fishing method employed on coral reefs in Southeast Asia. An encircling net is used with pounding devices, such as large stones fitted on ropes that are pounded onto the coral reefs. They can also consist of large heavy blocks of cement suspended above the sea by a crane fitted to the vessel. The pounding devices are repeatedly lowered into the area encircled by the net, smashing the coral into small fragments in order to scare the fish out of their coral refuges. The "crushing" effect on the coral heads has been described as having long-lasting and practically totally destructive effects.

ILLEGAL, UNREPORTED AND UNREGULATED FISHING

Illegal, unreported and unregulated fishing (IUU) is an international issue around the world. Industry observers believe IUU occurs in most fisheries, and accounts for up to 30% of total catches in some important fisheries.

Illegal fishing takes place when vessels or harvesters operate in violation of the laws of a fishery. This can apply to fisheries that are under the jurisdiction of a coastal state or to high seas fisheries regulated by regional fisheries management organisations (RFMO). According to the UN Food and Agriculture Organization (FAO), Fisheries and Aquaculture Department, illegal fishing has caused losses estimated at US$23 billion per year, with about 30% of illegal fishing in the world occurring in Indonesia alone.

Unreported fishing is fishing that has been unreported or misreported to the relevant national authority or RFMO, in contravention of applicable laws and regulations.

Unregulated fishing generally refers to fishing by vessels without nationality, vessels flying the flag of a country not party to the RFMO governing that fishing area or species on the high seas, or harvesting in unregulated areas.

The drivers behind illegal, unreported and unregulated (IUU) fishing are similar to those behind many other types of international environmental crime: pirate fishers have a strong economic incentive – many species of fish, particularly those that have been over-exploited and are thus in short supply, are of high financial value.

Such IUU activity may then show a high chance of success – i.e. a high rate of return – from the failure of governments to regulate adequately (e.g. inadequate coverage of international agreements), or to enforce national or international laws (e.g. because of lack of capacity, or poor levels of governance). A particular driver behind IUU fishing is the failure of a number of flag states to exercise effective regulation over ships on their registers – which in turn creates an incentive for ships to register under these flags of convenience.

Since no one reports catches made by pirates, their level of fishing cannot be accurately quantified.

Economic and Environmental impacts

One economic impact of illegal, unreported and unregulated (IUU) fishing on developing countries is the direct loss of the value of the catches that could be taken by local fishermen if the IUU fishing were not taking place.

These losses include a loss to GNP, and government revenues from landing fees, licence fees, and taxes payable by legal fishing operators. There are further indirect impacts in terms of loss of income and employment in related industries; losses in income will tend to reduce the consumer expenditures of families working in the fishing industry.

IUU harvests may be brought to market at a lower price as unfair competition to the same products from the regulated supply or as a mislabeled competing product. In either situation this illegal unregulated contribution to the market may lower the overall quality and price of products available, thus creating an economic burden on harvesters following the laws and regulations.

Illegal, unreported and unregulated (IUU) fishing can have a significant impact on the sustainability of both the targeted species and the ecosystem. Fishing generally has the capacity to damage fragile marine ecosystems and vulnerable species such as coral reefs, turtles and seabirds. In fact, all eight sea turtle species are now endangered, and illegal fishing and hunting are two major reasons for their destruction. Regulating legitimate fisheries is aimed at mitigating such impacts, but IUU fishers rarely comply with regulations. This may reduce future productivity and biodiversity and create imbalances in the ecosystem.

This may lead to reduced food security in communities heavily dependent on fish as a source of animal protein. IUU fishing can also lead to increased pressure on endangered fish species. IUU fishing can directly affect the population of fish species by increasing the number of fish caught within the population in spite of population management efforts by the international community. Indirectly, the substitution (mislabeling) of IUU caught fish for popular, but threatened or endangered species, increases the perceived supply of these species, thus decreasing the price and increasing the demand for the fish species.

Responsible Fishing Scheme

The Responsible Fishing Scheme is the only global standard that audits compliance on board fishing vessels, including ethical and welfare criteria. First launched in 2006 by Seafish to help fishing vessels demonstrate their commitment to a responsibly sourced catch, in January 2016 a revised scheme was launched to include the health and safety and welfare of crew on board. The new scheme has been re-developed in accordance with the requirements of internationally recognised standard ISO17065. Initially focused on vessels supplying the UK market, the scheme will be rolled out internationally over the next two years.

Seafish is the UK's authority on seafood. It was founded in 1981 by an Act of Parliament and aims to support all sectors of the seafood industry for a sustainable, profitable and socially responsible future. It is the only pan-industry body offering services to all parts of the industry, from the start of the supply chain at catching and aquaculture; through processing, importers, exporters and distributors of seafood right through to restaurants and retailers. Seafish is funded by a levy on the first sale of seafood landed in the UK. Its services are intended to support and improve the environmental sustainability, efficiency and cost-effectiveness of the industry, as well as promoting responsibly-sourced seafood.

Enforcement

Design for an environmental protection patrol vessel.

Illegal and unreported fishing (two of the three components of IUU fishing) essentially arise from a failure to adequately enforce existing national and international laws. There are, however, many factors underlying enforcement failure, including, notably, poor levels of national governance.

There are obvious problems with enforcing fisheries regulations on the high seas, including locating and apprehending the pirate ships, but solutions are available, chiefly through improved monitoring and surveillance systems.

Many fishermen are getting away with illegal, unreported, unregulated (IUU) ocean fishing due to the difficulty of monitoring every single fishing boat. This will begin to change with new technology that is able to watch and follow each boat from space consistently. This will be done through radio beacons, which all boats in the ocean are required to have. By implementing this new monitoring system, it will be impossible for fishermen to get away with IUU fishing in the future.

MSC systems are similarly of value within exclusive economic zones, including, for example, offshore patrols and licensing schemes.

Flag State Liability

Under the United Nations Convention on the Law of the Sea, states bear responsibility for the vessels (including fishing vessels) that fly their flag. While it is uncontroversial that individuals engaged in IUU fishing may be subject to legal sanctions, the extent to which flag states may be held liable under international law for the IUU fishing activities of their vessels is less clearly defined.

In March, 2013 the issue of flag state liability for IUU fishing was brought before the International Tribunal for the Law of the Sea in a request for an advisory opinion submitted by the Sub-Regional Fisheries Commission (SRFC) of West Africa. The SRFC asked the Tribunal to advise on the following four questions:

1. What are the obligations of the flag State in cases where illegal, unreported and unregulated (IUU) fishing activities are conducted within the Exclusive Economic Zone of third party States?
2. To what extent shall the flag State be held liable for IUU fishing activities conducted by vessels sailing under its flag?
3. Where a fishing license is issued to a vessel within the framework of an international agreement with the flag State or with an international agency, shall the State or international agency be held liable for the violation of the fisheries legislation of the coastal State by the vessel in question?
4. What are the rights and obligations of the coastal State in ensuring the sustainable management of shared stocks and stocks of common interest, especially the small pelagic species and tuna?

In its submissions to the Tribunal, the SRFC emphasized the severity of the IUU fishing problem in West Africa and the need for a clear regime of flag state responsibilities with respect to vessels engaged in this trade. The SRFC argued that its member states had been unable to mount successful prosecutions against IUU fishers following boardings due to a lack of support and cooperation from flag states.

A number of states and international organizations also made submissions adopting a range positions with respect to the four questions above. It was held by some states that the Tribunal lacked the advisory jurisdiction to respond to these questions. Both the government of Spain, a major contributor to the problem of IUU fishing, as well as the United Kingdom adopted such a view. A number of other states took the position that flag state responsibility for IUU fishing should take the form of a general responsibility to perform due diligence with respect to vessels and their activities, rather than an obligation to assist in prosecutions or another more substantive requirement.

In its Advisory Opinion issued on April 2, 2015, the Tribunal adopted the "due diligence" approach. The Tribunal found that flag states are under only a general obligation to take the necessary measures to ensure that their nationals and vessels flying their flag are not engaged in IUU fishing activities. This obligation may be satisfied by adhering to generally accepted international norms of fishing vessel regulation and complying with international treaties that indicate best practices. At the same time, the Tribunal found that in coastal waters the coastal state bears primary responsibility for preventing IUU fishing and not the flag state.

Political Processes

EU Action Plan

The EU helped draw up FAO's international plan of action to prevent, deter and eliminate IUU fishing, endorsed by the FAO Council in June 2001. The EU then developed its own plan to implement the commitments agreed at international level, and the European Commission's action plan for the eradication of IUU fishing was published in May 2002. It is intended to be implemented at four levels:

- At the EU level, more responsibility will be requested with regard to member state nationals acting under a flag of convenience. Moreover, market measures concerning fisheries products caught in violation of the international agreements will be adopted. In addition, information actions addressed to the fishing industry, consumers, and the public will be launched to raise their awareness.
- In the framework of Regional Fisheries Management Organisations, control and inspection plans would be adopted as well as specific conservation and management measures. In addition IUU vessels would be identified and monitored and their catches would be quantified.
- At the international level, concepts like genuine link would be defined, and a number of rights and obligations of the port state would be established. Moreover, the exchange of information on IUU activities and the international co-operation would be strengthened.
- In partnership with developing countries, the necessary means would be provided to enable them to effectively control fishing activities undertaken in waters under their jurisdiction.

High Seas Task Force

The High Seas Task Force comprises a group of fisheries ministers and international NGOs working together to develop an action plan designed to combat IUU fishing on the high seas.

Launched in 2003, the Task Force includes fisheries ministers from Australia, Canada, Chile, Namibia, New Zealand and the UK, together with the Earth Institute, IUCN-World Conservation Union, WWF International and the Marine Stewardship Council.

The goal of the Task Force is to set priorities among a series of practical proposals for confronting the challenge of IUU fishing on the high seas. A series of expert panels have been convened to identify the legal, economic, scientific and enforcement factors that permit IUU activity to thrive, and then determine key points of leverage that can brought to bear at national, regional, and global levels to minimise the incentives to carry out IUU fishing on the high seas. The completed action plan, published on 3 March 2006, will be placed by ministerial members of the Task Force directly in the hands of other ministers.

Regional Fisheries Management Organizations

Regional Fisheries Management Organisations (RFMOs) are affiliations of nations that co-ordinate efforts to manage fisheries in a particular region.

RFMOs may focus on certain species of fish (e.g. the Commission for the Conservation of Southern bluefin tuna) or have a wider remit related to living marine resources in general within a region (e.g. the Commission for the Conservation of Antarctic Marine Living Resources (CCAMLR)). This wide diversity of mandates and areas of application, and also effective implementation of regulations, opens up opportunities for IUU vessels.

UN High Seas Processes

The present system of high seas governance has evolved over a period of several hundred years, the end result being a patchwork quilt of measures in the form of binding and non-binding instruments with different geographical and legal reaches and different levels of participation.

Most legal instruments build on the foundation established by the UN Convention on the Law of the Sea, which was agreed in 1982 and entered into force in 1992.

The UN Fish Stocks Agreement, which entered into force in 2001, sets out principles for the conservation and management of fish stocks and establishes that such management must be based on the precautionary approach and the best available scientific information. The Agreement provides a framework for cooperation on conservation and management, but since only about a third of the parties to the Law of the Sea Convention have ratified it, its impact is inevitably limited.

The UN Food and Agriculture Organisation (FAO) carries out much of the technical work on international fisheries management, and provides a forum for the negotiation of agreements and codes of conduct. In 1995 the FAO agreed its Code of Conduct for Responsible Fisheries to promote long-term sustainable management.

In 2001, the FAO adopted the International Plan of Action (IPOA) on IUU Fishing. The aim of this voluntary instrument is to prevent, deter and eliminate IUU fishing by providing all states with comprehensive, effective and transparent measures by which to act, including through appropriate regional fisheries management organisations established in accordance with international law.

The FAO Compliance Agreement, which entered into force in 2003, is designed to close a major loophole in international fisheries management, that of the circumvention of fisheries regulations by 're-flagging' vessels under the flags of states that are unable or unwilling to enforce such measures.

In 2009, the FAO brokered the Agreement on Port State Measures to Prevent, Deter and Eliminate Illegal, Unreported and Unregulated Fishing, which entered into force in 2016. The agreement closes ports to vessels suspected of illegal fishing. The treaty requires that fishing vessels request permission to dock and inform the port of the details of its fishing operations. Permission to dock can be denied if unregulated fishing was occurring. The measure is intended to block illegally caught fish from entering the marketplace. Other measures in the treaty include inspections of equipment, paperwork, catches, and ship's records. Though the treaty does not compel countries to apply these measures to ships under their own flags, they may choose to do so under the agreement.

ENVIRONMENTAL IMPACT OF FISHING

Humans have relied on the ocean for subsistence by harvesting its abundance of fish. In recent decades, new technologies have allowed humans to remove fish from the ocean on a massive scale to supply Earth's burgeoning population. Unfortunately, there are many negative environmental consequences to these practices and overfishing has been identified as a primary cause of ecosystem collapse in many aquatic systems.

Habitat Destruction

In addition to removing an increasingly large number of fish from the ocean, many industrial fishing practices also destroy aquatic habitat. Dredging is a practice commonly used to harvest clams and employs a large metal scoop that drags along the seafloor to pick them up. The process also churns up sediments along the seafloor, causing them to become suspended in the water column, decreasing water quality. The practice can also dig up burrowing worms from the sediments. These animals are important because their burrows increase contact between sediments and the water. This returns nutrients to the water, where they are used by microorganisms in nutrient cycling. Without these burrowing animals, waters along the seafloor can become depleted of oxygen and uninhabitable.

Seafloor trawling, in which equipment is pulled across the seafloor to catch bottom-dwelling fish, decreases the biomass and production of benthic species. The practice also practice destroys corals, oysters and sponges that form productive marine habitats. The impacts of the destruction can be far-reaching. Oyster reefs, for example, have been decimated in many places from trawling. Without these important filter feeders, coastal waters can suffer from eutrophication, in which there are too many nutrients in the water. This in turn causes harmful algal blooms that can cause widespread die-offs of marine life.

In tropical regions, people sometimes employ blast fishing. In this method, fishermen light sticks of dynamite and toss them into the water. The explosion stuns nearby fish and can make their swim bladders rupture, causing them to float to the surface for easy capture. With up to $2,000 worth of fish being caught from a single blast, the process can be lucrative for fishermen, but destroys coral reef habitat in the process. Like oyster beds, coral reefs are productive habitats that serve as nurseries for many fishes. With these important developmental habitats gone, stocks of many species of fish can rapidly decline. Many of these species are also commercially valuable, so although the short-term payoff is large, the practice can actually be harmful to fishermen in the long run. Furthermore, people also lose the valuable ecosystem services provided by coral reefs, such as coastal protection and revenue through tourism.

Derelict Fishing Gear

Some kinds of fishing gear can be even more destructive when they become lost or forgotten in the water because they continue to catch animals, a phenomenon known as "ghost fishing." This is particularly wasteful and destructive because the gear can ensnare tons of animals that aren't being harvested or used in any way. Fishing piers can be sites of ghost fishing as lures and lines become wrapped around pilings, where animals swimming by become trapped. Fish are not the only victims, however, as birds that dive into the water for prey can also get caught in the lines when they enter the water.

Diamondback terrapins provide a case study of how ghost fishing can impact animal populations. These turtles inhabit salt marshes along the East Coast where people also fish for blue crabs. Blue crabs are caught using a crab pot, a metal cage that is dropped to the floor of the marsh and tied to a buoy that floats along the surface. If the buoy becomes detached, fishermen may not be able to find the gear, and it becomes a "ghost pot." Crab pots have tiny openings that allow crabs to enter, but are constructed in a way that makes it difficult for them to get back out. Terrapins also swim into the pots, attracted by the bait in the middle. Because they are social animals, when several turtles are in a pot, it often draws in other individuals, leading the ghost pot to catch more and more turtles over time. A single ghost pot was discovered in Georgia that contained more than 130 deceased turtles. This fishing gear is thought to pose a major threat to many populations of diamondback terrapin.

In addition to direct mortality, ghost pots can also alter population demographics. This is because female terrapins are larger than males and females often cannot fit into the openings in crab pots. As a result, the pots trap males and younger turtles at a higher rate, shifting the population to older turtles and females. This biased sex ratio, along with the removal of many individuals before they reach sexual maturity, can further exacerbate declines already faced by these populations.

Bycatch

While many fishing practices can be extremely adept at harvesting fish, they also often incidentally catch non-target species, known as bycatch. Longline fishing is one of the major bycatch culprits. This practice is often employed in the open ocean and consists of miles of lines off of which thousands of baited hooks are attached and a single long line can have more than 2,500 hooks. Sea turtles are especially prone to capture in longlines since they must reach the surface of the ocean to breathe. When they go for the bait on the lines, the hooks become lodged in their mouths and they drown because they cannot reach the surface. Even when they are not attracted by the bait, some turtles are so large that the lines can become wrapped around their flippers as they swim through them. Marine mammals often become trapped in the lines as well. Birds like albatross that fly over the open seas are attracted by the bait as the lines are put into the water. Once hooked on the line, they are ensnared in the gear and soon drown.

Other fishing practices can also incur large amounts of bycatch. Trawling can drown turtles that get caught in nets. Gillnets are staked to the ocean floor and entrap many animals unintentionally. The collective toll from these practices is staggering. Over the past twenty years, an estimated 85,000 sea turtles have been killed as bycatch. Additionally, an estimated 300,000 marine mammals, 160,000 albatross and 3 million sharks are lost to bycatch from fishing practices each year. This high rate of mortality is not sustainable for these animal populations. Animals like albatrosses and sea turtles that are long-lived and slow to reach maturity are particularly impacted by these threats and many populations have declined precipitously over the past few decades.

As with the other impacts from fishing, consequences of bycatch are often far-reaching as species become functionally extinct in many areas. For example, leatherback turtles are major predators of jellyfish, capable of consuming more than 600 jellyfish in a single day. With their major predators gone, there have been jellyfish population booms in some parts of the ocean. This has made the waters dangerous for swimmers in some places that rely on tourism as a revenue source and can also prevent the recovery of depleted fish stocks. Green sea turtles and manatees, which are commonly victims of bycatch, also demonstrate the ecological complexities that can occur after the removal of important functional species. These animals are herbivores that graze in seagrass beds. In these habitats they are important because their constant grazing keeps the grass at a healthy level, contributing to nutrient cycling and preventing disease outbreaks from spreading through the grass beds. Without these grazers, many seagrass ecosystems have suffered from large seagrass die-offs. When this happens, biodiversity declines as these habitats can no longer support a wide range of marine life.

FISHERIES AND CLIMATE CHANGE

The plethora of effects related to climate change involving, besides climate, oceans, coasts and freshwater ecosystems, are bound to affect fisheries and habitats together with the composition and location of production and will have major impacts on aquaculture productivity and security.

Climate change imperils the structure and function of already stressed coastal aquatic ecosystems. Estuaries, coral reefs, mangroves and sea grass beds are critical for production of wild fish. In freshwater systems, ecosystem health and productivity is linked to water quality and flow and the health of wetlands.

Small wild fish like anchovies and sardines are sensitive to changes in ocean conditions. They are found in schools in the ocean and are also processed into fishmeal used to feed other fish (aquaculture) as well as poultry and pigs.

Though precise consequences cannot yet be forecast, climate change is likely to affect fisheries and aquaculture, their dependent communities and related economic activities along three main pathways:

- Indirect wider socio-economic effects (e.g. fresh water use conflicts affect all food production systems, adaptation and mitigation strategies in other sectors impact aquatic systems in general or fisheries and aquaculture directly).
- Biological and ecological responses to physical changes (e.g. productivity, species abundance, ecosystem stability, stock locations, pathogen levels and impacts).
- Direct physical effects (e.g. sea level change, flooding, storm impacts).

Fishers, fish farmers and coastal inhabitants will bear the full force of these impacts through less stable livelihoods, changes in the availability and quality of fish for food, and rising risks to their health, safety and homes. Many fisheries-dependent communities already live a precarious and vulnerable existence because of poverty, lack of social services and essential infrastructure. The fragility of these communities is further undermined by overexploited fishery resources and degraded ecosystems. The implications of climate change for food security and livelihoods in small island states and many developing countries are profound.

MARINE POLLUTION

Marine pollution has been an ever-present problem since the advent of large-scale agricultural activity and industrialization. However, significant laws and regulations at an international level to tackle the problem came only in the mid-twentieth century. During United Nations Conventions on the Law of the Sea in the early 1950s, the various stakeholders come together to deliberate and formulate laws pertaining to marine pollution. Till mid-twentieth century the majority of the scientists maintained that oceans were vast enough to be able to dilute the amount of pollution being drained into them, thus, considering pollution harmless to the marine life.

Causes of Marine Pollution

The marine environment becomes polluted and contaminated through various sources and forms. Major sources of marine pollution are the inflow of chemicals, solid waste, discharge of radioactive elements, industrial and agricultural effluents, man-made sedimentation, oil spills, and many such factors. The majority portion of the marine pollution comes from the land that contributes to 80 percent of the marine pollution, air pollution also carries pesticides from farms and dust into the marine waters. Air and land pollution is a major contributor to the growing marine pollution that is not only hampering the aquatic ecology but also affecting the life on land. The non-point sources like wind-blown debris, agricultural runoff, and dust become the major source of pollution. Apart from these, factors like land runoff, direct discharge, atmospheric pollution, pollution caused by ships, and deep sea mining of natural resources contribute heavily.

Types of Marine Pollution

Eutrophication

When there is an excess of chemical nutrients mainly nitrates and phosphates in the water, it leads to eutrophication or nutrient pollution. Eutrophication decreases the level of oxygen, reduces the quality of water, makes the water inhabitable for fish, affects the breeding process within the marine life and increases the primary productivity of the marine ecosystem.

Acidification

Oceans act as a natural reservoir for absorbing the carbon dioxide from the Earth's atmosphere. But, due to rising level of carbon dioxide in the atmosphere, the oceans across the world are becoming acidic in nature, as a consequence, it leads to acidification of oceans. Researches and scientists have not been able to uncover the potential damage ocean acidification may have on the Earth's atmosphere. But, there is a strong concern that acidification might lead to dissolution of calcium carbonate structures, that can affect the shell formation in shellfish and also the corals.

Toxins

There are persistent toxins that do not get dissolved or disintegrate with the marine ecosystem rapidly. Toxins such as pesticides, DDT, PCBs, furans, TBT, radioactive waste, phenols, and dioxins get accumulated in the tissue cells of the marine lifeforms and lead to bioaccumulation hampering the life underwater and sometimes leads to a mutation in aquatic life forms.

Plastics

The ever-growing dependence of human population on plastic has filled the oceans and the land, it consists of 80 percent of the debris found in the oceans. Plastic dumped and found in the oceans are dangerous for the marine life forms and wildlife, as sometimes it strangles and chokes them to death. The rising levels of plastic dumps found in the oceans are suffocating, ingesting, and entangling the life underwater as well as above it.

Effects of Marine Pollution

The contamination of water by excessive nutrients is known as nutrient pollution, a type of water pollution that affects the life under water. When excess nutrients like nitrates or phosphates get dissolved with the water it causes the eutrophication of surface waters, as it stimulates the growth of algae due to excess nutrients. Most of Benthic animals and plankton are either filter feeders or deposit feeders take up the tiny particles that adhere to potentially toxic chemicals. In the ocean food chains, such toxins get concentrated upward. This makes estuaries anoxic as many particles combine chemically depletive of oxygen.

When the marine ecosystem absorbs the pesticides, they are incorporated into the food webs of the marine ecosystem. After getting dissolved in the marine food webs, these harmful pesticides causes mutations, and also results in diseases, which can damage the entire food web and cause harm to the humans. When toxic metals are dumped or flown into the oceans through drains, it engulfs within the marine food webs. It affects the biochemistry, reproduction process, can affect the tissue matter These can cause a change to tissue matter, biochemistry, behavior, reproduction, and suppress and alter the marine life's growth. Marine toxins can be transferred to several animals feeding on the fish or fish hydrolysate as a meal, toxins are then transferred to dairy products and meat of these affected land animals.

Steps to Prevent Marine Pollution

- Stop using plastic and littering garbage as they not only choke up the drains but also releases into the oceans.
- Ensure that chemicals mentioned above are not used anywhere near the streams of water and try cutting down on the usage of such chemicals.
- For farmers, they need to switch from chemical fertilizers and pesticides and move towards the usage of organic farming methods.
- Use public transport and reduce the carbon footprint by taking small and substantial measures that will not help in reducing the pollution from the environment but will ensure a safe and healthy future for the upcoming generations.
- Prevent from any oil or chemical spill in the oceans and if in case there is an oil or chemical spill near you volunteer and help in cleaning out the ocean water.
- Volunteer or initiate beach clean up activities and spread awareness about the same in the nearby vicinity.

References

- Pauly, daniel; froese, rainer (2012). "comments on fao's state of fisheries and aquaculture, or 'sofia 2010'". Marine policy. 36 (3): 746–752. Doi:10.1016/j.marpol.2011.10.021

- Pulse-fishing, dossiers: wur.nl, Retrieved 8 January, 2019
- bassan, janine (may–june 2011). "not all seafood is equal". South african journal of science. 107 (5/6): 8–10. Doi:10.4102/sajs.v107i5/6.718
- scales, helen (29 march 2007). "shark declines threaten shellfish stocks, study says". National geographic news. Retrieved 2012-05-01
- Sustainable-fishing, encyclopedia: nationalgeographic.org, Retrieved 5 February, 2019
- Environmental-consequences-fishing-practices: environmentalscience.org, Retrieved 31 March, 2019
- united nations general assembly session 59 verbatim report 56. A/59/pv.56 page 4. Ms. Kimball international union for the conservation of nature and natural resources 17 november 2004. Retrieved 2009-05-09
- Marine-pollution-causes-types-effects-prevention, society, my-india: mapsofindia.com, Retrieved 16 January, 2019

PERMISSIONS

All chapters in this book are published with permission under the Creative Commons Attribution Share Alike License or equivalent. Every chapter published in this book has been scrutinized by our experts. Their significance has been extensively debated. The topics covered herein carry significant information for a comprehensive understanding. They may even be implemented as practical applications or may be referred to as a beginning point for further studies.

We would like to thank the editorial team for lending their expertise to make the book truly unique. They have played a crucial role in the development of this book. Without their invaluable contributions this book wouldn't have been possible. They have made vital efforts to compile up to date information on the varied aspects of this subject to make this book a valuable addition to the collection of many professionals and students.

This book was conceptualized with the vision of imparting up-to-date and integrated information in this field. To ensure the same, a matchless editorial board was set up. Every individual on the board went through rigorous rounds of assessment to prove their worth. After which they invested a large part of their time researching and compiling the most relevant data for our readers.

The editorial board has been involved in producing this book since its inception. They have spent rigorous hours researching and exploring the diverse topics which have resulted in the successful publishing of this book. They have passed on their knowledge of decades through this book. To expedite this challenging task, the publisher supported the team at every step. A small team of assistant editors was also appointed to further simplify the editing procedure and attain best results for the readers.

Apart from the editorial board, the designing team has also invested a significant amount of their time in understanding the subject and creating the most relevant covers. They scrutinized every image to scout for the most suitable representation of the subject and create an appropriate cover for the book.

The publishing team has been an ardent support to the editorial, designing and production team. Their endless efforts to recruit the best for this project, has resulted in the accomplishment of this book. They are a veteran in the field of academics and their pool of knowledge is as vast as their experience in printing. Their expertise and guidance has proved useful at every step. Their uncompromising quality standards have made this book an exceptional effort. Their encouragement from time to time has been an inspiration for everyone.

The publisher and the editorial board hope that this book will prove to be a valuable piece of knowledge for students, practitioners and scholars across the globe.

INDEX